# Springer Theses

Recognizing Outstanding Ph.D. Research

## Aims and Scope

The series "Springer Theses" brings together a selection of the very best Ph.D. theses from around the world and across the physical sciences. Nominated and endorsed by two recognized specialists, each published volume has been selected for its scientific excellence and the high impact of its contents for the pertinent field of research. For greater accessibility to non-specialists, the published versions include an extended introduction, as well as a foreword by the student's supervisor explaining the special relevance of the work for the field. As a whole, the series will provide a valuable resource both for newcomers to the research fields described, and for other scientists seeking detailed background information on special questions. Finally, it provides an accredited documentation of the valuable contributions made by today's younger generation of scientists.

## Theses are accepted into the series by invited nomination only and must fulfill all of the following criteria

- They must be written in good English.
- The topic should fall within the confines of Chemistry, Physics, Earth Sciences, Engineering and related interdisciplinary fields such as Materials, Nanoscience, Chemical Engineering, Complex Systems and Biophysics.
- The work reported in the thesis must represent a significant scientific advance.
- If the thesis includes previously published material, permission to reproduce this must be gained from the respective copyright holder.
- They must have been examined and passed during the 12 months prior to nomination.
- Each thesis should include a foreword by the supervisor outlining the significance of its content.
- The theses should have a clearly defined structure including an introduction accessible to scientists not expert in that particular field.

More information about this series at http://www.springer.com/series/8790

Valentin Knünz

# Measurement of Quarkonium Polarization to Probe QCD at the LHC

Doctoral Thesis accepted by
the Vienna University of Technology, Austria

 Springer

*Author*
Dr. Valentin Knünz
Experimental Physics Department
CERN
Geneva
Switzerland

*Supervisor*
Dr. Claudia-Elisabeth Wulz
Institute of High Energy Physics
    of the Austrian Academy of Sciences
Vienna
Austria

ISSN 2190-5053          ISSN 2190-5061   (electronic)
Springer Theses
ISBN 978-3-319-84278-3          ISBN 978-3-319-49935-2   (eBook)
DOI 10.1007/978-3-319-49935-2

Printed on acid-free paper

This Springer imprint is published by Springer Nature
The registered company is Springer International Publishing AG
The registered company address is: Gewerbestrasse 11, 6330 Cham, Switzerland

*Physics is not the most important thing.*
*Love is.*

Richard Feynman

# Supervisor's Foreword

The strong interaction is one of the fundamental forces in the universe. It binds quarks into hadrons and is responsible for holding atomic nuclei together. Quantum chromodynamics (QCD) is the modern field theory that describes many aspects of the strong interaction. However, open questions remain, such as the exact nature of the hadronization process. Quarkonia, bound states of charm or beauty quark–antiquark pairs, are particularly suited to illuminate these questions, as the masses of the involved quarks are large, and consequently, their velocities are small, which makes it easier to describe their interactions. Many experiments have therefore studied quarkonia. However, measurements were inconclusive and in partial contradiction to theoretical modeling both with respect to production cross sections and polarizations. Valentin Knünz's thesis represents a major breakthrough to solve the long-standing "quarkonium puzzle", using data of the CMS experiment at CERN's Large Hadron Collider (LHC) that extend to unprecedentedly high transverse momenta. The extraordinary relevance of Valentin's thesis is also reflected in an article of the CERN Courier published in February 2015, with the title "CMS heads toward solving a decades-long quarkonium puzzle" [1].

The analyses performed by Valentin have enormously contributed to the resolution of inconsistencies and to the consolidation of newly achieved, rather astonishing results. On the one hand, they were among the first measurements of quarkonium polarization at the LHC, and on the other, they contributed to the physical interpretation in the framework of a global analysis of quarkonia data.

Valentin developed innovative techniques for the polarization measurement, which corroborate the results through internal consistency checks that were never made before. Furthermore, his thesis excels through his careful determination of background, acceptances, and systematic errors, largely using experimental data. It must be remarked that polarization measurements of quarkonia are lengthy and difficult, which is also reflected by the initially contradictory results obtained by the experiments CDF and D0 at the Tevatron.

In his thesis, Valentin established the first polarization measurements at the LHC for Upsilon states and clearly demonstrated a solution to the quarkonium polarization puzzle, in particular by determining the transverse momentum range in which the non-relativistic QCD theory (NRQCD) is applicable. He was therefore chosen to present the results for the first time at the International Conference of High Energy Physics 2012 in Melbourne. Furthermore, Valentin supported the analysis of the polarization of J/$\psi$ particles. The final results, at 7 TeV LHC center-of-mass energy, were published in [2, 3], which are highly cited. A further highly cited publication [4] on the phenomenological interpretation of the measurements, written with only four co-authors, is a spin-off of this thesis, which shows Valentin's great theoretical knowledge. After the polarization measurements, which would already have represented sufficient material for a doctoral thesis, Valentin dedicated himself to a phenomenological study of polarization and cross-sectional measurements of ATLAS, CMS, and LHCb. All data were simultaneously and consistently compared to a superposition of color singlet and octet terms in the framework of the NRQCD model. This work led to the determination of long-distance matrix elements, which are directly related to the formation of bound states in the non-perturbative regime. None of the previous fits was able to correctly reproduce the correlations between the measured cross sections and polarizations. Within NRQCD, it was initially assumed that quarkonia were produced as quark–antiquark pairs in the color octet $^3S_1$ configuration, corresponding to the quantum numbers of a gluon. This would have resulted in a transverse polarization of quarkonia with large transverse momenta, just as it is the case for high-energy gluons. Instead, Valentin showed for the first time that the data can be correctly reproduced with the assumption of an initially colored spherically symmetric $^1S_0$ quark–antiquark state. After losing its color by emitting soft gluons and transforming into the observed quarkonium, it decays isotropically. This is precisely what the CMS data show.

Apart from the CMS Thesis Award, Valentin's work won two other prizes—the Victor Hess Prize of the Nuclear and Particle Physics Division of the Austrian Physical Society and the Special Prize for Science attributed by the Austrian region of Vorarlberg.

Geneva                                                                    Dr. Claudia-Elisabeth Wulz
July 2016

# References

1. Cerncourier.com/cws/article/cern/60147
2. CMS Collaboration (2013) Measurement of the $\Upsilon(1S)$, $\Upsilon(2S)$ and $\Upsilon(3S)$ polarizations in $pp$ collisions at $\sqrt{s} = 7$ TeV. Phys Rev Lett 110:081802
3. CMS Collaboration (2013) Measurement of the prompt $J/\psi$ and $\psi(2S)$ polarizations in $pp$ collisions at $\sqrt{s} = 7$ TeV. Phys Lett B 727:381
4. Faccioli P, Knünz V, Lourenço C, Seixas J, Wöhri H (2014) Quarkonium production in the LHC era: a polarized perspective. Phys Lett B 736:98

# Abstract

With the first proton–proton collisions in the Large Hadron Collider (LHC) at CERN in 2010, a new era in high energy physics has been initiated. The data collected by the various experiments open up the possibility to study standard model processes with high precision, in new areas of phase space. The LHC provides excellent conditions for studies of quarkonium production, due to the high quarkonium production rates given the high center-of-mass energy and high instantaneous luminosity of the colliding proton beams. Studies of the production of heavy quarkonium mesons—bound states of a heavy quark and its respective antiquark—are very important to improve our understanding of hadron formation. Until quite recently, experimental and phenomenological efforts have not resulted in a satisfactory overall picture of quarkonium production cross sections and quarkonium polarizations.

The Compact Muon Solenoid (CMS) detector is ideally suited to study quarkonium production in the experimentally very clean dimuon decay channel, up to considerably higher values of transverse momentum than accessible in the previous experiments. The scope of this thesis was to describe in detail the measurements of the polarizations of the $\Upsilon(nS)$ bottomonium states and (in less detail) of the $\psi(nS)$ charmonium states, based on a dimuon data sample collected with the CMS detector in proton–proton collisions at a center-of-mass energy of 7 TeV. Surprisingly, no significant polarizations were found in any of the studied quarkonium states, in none of the studied reference frames, nor in a frame-independent analysis. From an experimental point of view, these results, together with recent results from other experiments, clarify the confusing picture originating from previous measurements, which were plagued by experimental ambiguities and inconsistencies.

The currently most favored approach to model and understand quarkonium production is non-relativistic quantum chromodynamics (NRQCD), a QCD-inspired model which allows color-octet pre-resonant quark–antiquark states to contribute to quarkonium bound state formation. The measurements obtained as a result of this work, together with other LHC measurements in the field of quarkonium production, are interpreted with an original phenomenological approach within the theoretical framework of NRQCD, guided by the observation of a few general features of the data, and corroborated by a detailed study of the quarkonium production cross section and polarization observables. This phenomenological analysis leads to a coherent picture of quarkonium production cross sections and polarizations within a simple model, dominated by one single color-octet production mechanism. These findings provide new insight in the dynamics of heavy quarkonium production at the LHC, an important step toward a satisfactory understanding of hadron formation within the standard model.

# Acknowledgements

This thesis represents the end of a five-year journey, beginning with the research for my master's degree. This work would not have been possible without the immense support of many people along the years whom I want to thank here.

First, I want to thank Chris Fabjan for awakening my interest in particle physics, for giving me the chance to start with my first research project at HEPHY, and for his interest in my work and his continuous support throughout the years; Jozko Strauss, for his welcome advice on diverse matters regarding my work, for his initiative to deepen my knowledge about particle physics and to strengthen the collaboration with theorists, and for developing the grant proposal that made this thesis possible; Claudia-Elisabeth Wulz, for her willingness to take over the supervision of my Ph.D. thesis in the middle of my research work and for her experienced guidance and professional support; André Hoang, for reviewing this thesis.

I received great support from many HEPHY colleagues. First, I want to thank my colleague Ilse Krätschmer for fostering a very helpful and friendly working atmosphere throughout the years; Wolfgang Adam, for his support in a large variety of issues, including strategic planning, solving technical problems, and frequent comments on and crucial inputs for my work; Rudi Frühwirth, for sharing his vast knowledge about statistical methods, his regular involvement in my early work, and his general advice. I further want to mention only a small fraction of the colleagues at HEPHY whose support was most important during these years, providing scientific input through interesting discussions, helping to solve coding problems, as well as institutional support: I want to thank Herbert Rohringer, Wolfgang Kiesenhofer, Wolfgang Waltenberger, Robert Schöfbeck, and Jochen Schieck.

My gratitude to Carlos Lourenço cannot be stressed enough: Most thoughts mentioned in this thesis are based on almost daily discussions with Carlos, making him my in-official co-supervisor and mentor. Most of my presentations, abstracts, proceedings, analysis notes, as well as this thesis were meticulously proofread in full by Carlos and profited from his maddening attention to detail, improving my scientific output from start to fin(n)ish. I further want to thank Carlos for his reliability when I needed help, and for his advice regarding my career planning, both on a professional and personal level.

I further want to thank the additional members of the "quarkonium polarization team", i.e., Hermine Wöhri, Pietro Faccioli, and João Seixas, for their endless efforts to make me understand the details of quarkonium production and polarization, for all the help regarding my work, for including me in the research leading to our latest published results, for many enjoyable meetings in Vienna, CERN, and Lisbon, and, finally, for introducing me to the joys of eating pastéis de nata.

I want to thank all proofreaders (which are all mentioned above for different reasons) for carefully reading my thesis and providing valuable comments, considerably improving the quality of this document. Within the CMS collaboration, I would like to thank all not yet mentioned members of the BPH group who helped to improve my work with their reviews and (in most cases) very valuable comments. This thesis, as well as the presentation of the corresponding results in numerous international conferences, was made possible by the FWF project P 24167-N16.

On a personal note, even though unrelated to the content of the thesis, I want to thank my friends and family, for their unconditional support during these years in high energy physics, for their honest and encouraging advice, and for helping to put things into perspective.

Finally, I want to thank Regina, the most significant person in my life, for her love, for being there whenever I need her, for the acceptance, understanding, and support in demanding times during and especially at the end of these Ph.D. years, and, most importantly, for the wonderful shared adventure(s) of bringing a new life into this world.

# Contents

| | | | |
|---|---|---|---:|
| **1** | **Introduction** | | 1 |
| | References | | 5 |
| **2** | **Quarkonium Physics** | | 9 |
| | 2.1 | Introduction | 9 |
| | 2.2 | Quarkonium Production | 13 |
| | | 2.2.1    Non-relativistic QCD Factorization | 14 |
| | | 2.2.2    Quarkonium Production in the Pre-LHC Era | 20 |
| | 2.3 | Quarkonium Polarization | 23 |
| | | 2.3.1    General Considerations | 24 |
| | | 2.3.2    Frame-Invariant Formalism | 27 |
| | | 2.3.3    Ambiguity of Pre-LHC Quarkonium Polarization Measurements | 29 |
| | | 2.3.4    Polarization of the $\chi$ States | 29 |
| | 2.4 | Quarkonium Physics Summary | 31 |
| | References | | 31 |
| **3** | **Experimental Setup** | | 35 |
| | 3.1 | The Large Hadron Collider | 35 |
| | | 3.1.1    The Machine | 35 |
| | | 3.1.2    Physics at the LHC | 38 |
| | 3.2 | The Compact Muon Solenoid Experiment | 40 |
| | | 3.2.1    Design | 40 |
| | | 3.2.2    Tracking Detectors | 42 |
| | | 3.2.3    Muon Detectors | 44 |
| | | 3.2.4    Trigger and Data Acquisition Systems | 47 |
| | | 3.2.5    Offline Track and Muon Reconstruction | 55 |
| | | 3.2.6    Photon Conversion Reconstruction | 60 |
| | | 3.2.7    Quarkonium Reconstruction Performance | 69 |
| | 3.3 | Experimental Setup Summary | 73 |
| | References | | 74 |

**4   Data Analysis**. . . . . . . . . . . . . . . . . . . . . . . . . . . . . . . . . . . . . . . . . .   77
    4.1   Analysis Strategy . . . . . . . . . . . . . . . . . . . . . . . . . . . . . . . . . . . . .   77
          4.1.1   The Polarization Extraction Framework . . . . . . . . . . . . . . .   79
          4.1.2   Background Subtraction . . . . . . . . . . . . . . . . . . . . . . . . . .   80
          4.1.3   Posterior Probability Density of the Anisotropy
                  Parameters. . . . . . . . . . . . . . . . . . . . . . . . . . . . . . . . . . . . .   81
          4.1.4   Extraction of the Results. . . . . . . . . . . . . . . . . . . . . . . . . .   85
          4.1.5   Validation of the Framework . . . . . . . . . . . . . . . . . . . . . .   88
    4.2   Measurement of the $\Upsilon(nS)$ Polarizations . . . . . . . . . . . . . . .   91
          4.2.1   $\Upsilon(nS)$ Data Processing and Event Selection . . . . . . . . . . .   91
          4.2.2   $\Upsilon(nS)$ Efficiencies. . . . . . . . . . . . . . . . . . . . . . . . . . . . .   92
          4.2.3   $\Upsilon(nS)$ Mass Distribution. . . . . . . . . . . . . . . . . . . . . . . . .   97
          4.2.4   Determination of the $\Upsilon(nS)$ Background Model . . . . . . . .  101
          4.2.5   Systematic Uncertainties . . . . . . . . . . . . . . . . . . . . . . . . . .  104
          4.2.6   Results . . . . . . . . . . . . . . . . . . . . . . . . . . . . . . . . . . . . . . .  106
    4.3   Measurement of the Prompt $\psi(nS)$ Polarizations . . . . . . . . . . . . .  109
          4.3.1   $\psi(nS)$ Data Processing and Event Selection . . . . . . . . . . .  109
          4.3.2   $\psi(nS)$ Efficiencies . . . . . . . . . . . . . . . . . . . . . . . . . . . . .  109
          4.3.3   $\psi(nS)$ Mass and Lifetime Distributions. . . . . . . . . . . . . . .  111
          4.3.4   Determination of the $\psi(nS)$ Background Model . . . . . . . .  115
          4.3.5   Summary of the Systematic Uncertainties . . . . . . . . . . . . .  116
          4.3.6   Results . . . . . . . . . . . . . . . . . . . . . . . . . . . . . . . . . . . . . . .  118
    4.4   Data Analysis Summary . . . . . . . . . . . . . . . . . . . . . . . . . . . . . . . .  119
    References. . . . . . . . . . . . . . . . . . . . . . . . . . . . . . . . . . . . . . . . . . . . . .  120

**5   Discussion of Results**. . . . . . . . . . . . . . . . . . . . . . . . . . . . . . . . . . . .  123
    5.1   Quarkonium Production Data at the LHC. . . . . . . . . . . . . . . . . . . .  123
          5.1.1   Cross Section Measurements. . . . . . . . . . . . . . . . . . . . . . .  123
          5.1.2   Polarization Measurements . . . . . . . . . . . . . . . . . . . . . . . .  126
    5.2   NRQCD Analyses Review . . . . . . . . . . . . . . . . . . . . . . . . . . . . . . .  129
    5.3   A Data-Driven Perspective . . . . . . . . . . . . . . . . . . . . . . . . . . . . . .  137
          5.3.1   Theory Ingredients . . . . . . . . . . . . . . . . . . . . . . . . . . . . . .  138
          5.3.2   Fitting Method . . . . . . . . . . . . . . . . . . . . . . . . . . . . . . . . .  141
          5.3.3   Kinematic Domain Scan . . . . . . . . . . . . . . . . . . . . . . . . . .  143
          5.3.4   Results and Predictions . . . . . . . . . . . . . . . . . . . . . . . . . . .  146
          5.3.5   Comparison with Other NRQCD Analyses . . . . . . . . . . . . .  151
          5.3.6   Conclusions. . . . . . . . . . . . . . . . . . . . . . . . . . . . . . . . . . . .  152
    5.4   Results Summary . . . . . . . . . . . . . . . . . . . . . . . . . . . . . . . . . . . . .  153
    References. . . . . . . . . . . . . . . . . . . . . . . . . . . . . . . . . . . . . . . . . . . . . .  154

**6   Conclusions** . . . . . . . . . . . . . . . . . . . . . . . . . . . . . . . . . . . . . . . . . . .  157
    6.1   Thesis Summary. . . . . . . . . . . . . . . . . . . . . . . . . . . . . . . . . . . . . .  157
    6.2   Outlook . . . . . . . . . . . . . . . . . . . . . . . . . . . . . . . . . . . . . . . . . . . .  159

**Curriculum Vitae** . . . . . . . . . . . . . . . . . . . . . . . . . . . . . . . . . . . . . . . . . .  161

# Abbreviations

| | |
|---|---|
| ALICE | A Large Ion Collider Experiment |
| AOD | Analysis Object Data |
| ATLAS | A Toroidal LHC ApparatuS |
| BK | Butenschön, Kniehl |
| BPH | Beauty-PHysics |
| BPix | Barrel Pixel Detector |
| BS | Beam Spot |
| BSM | Beyond the Standard Model |
| CB | Crystal Ball |
| CDF | Collision Detector Fermilab |
| CEM | Color-Evaporation Model |
| CERN | Organisation Européenne pour la Recherche Nucléaire |
| CL | Confidence Level |
| CM | Center of Mass |
| CMS | Compact Muon Solenoid |
| CMSSW | CMS SoftWare |
| CMSWZ | Chao, Ma, Shao, Wang, Zhang |
| CO | Color Octet |
| CS | Color Singlet |
| CSC | Cathode Strip Chamber |
| CSM | Color-Singlet Model |
| CTF | Combinatorial Track Finder |
| CV | Conversion Vertex |
| DAQ | Data AcQuisition |
| DCA | Distance of Closest Approach |
| DQM | Data Quality Monitoring |
| DT | Drift Tube |
| ECAL | Electromagnetic CALorimeter |
| *ep* | electron–proton |
| FSR | Final State Radiation |

| FP | Fraction Parameter |
|---|---|
| FPix | Forward Pixel Detector |
| GWWZ | Gong, Wan, Wang, Zhang |
| HCAL | Hadron CALorimeter |
| HEPHY | Institut für Hochenergiephysik/Institute of High Energy Physics |
| HERA | Hadron Elektron Ring Anlage |
| HI | Heavy Ion |
| HLT | High-Level Trigger |
| HPD | High Posterior Density |
| HX | Helicity |
| IP | Interaction Point |
| ISP | InterSection Point |
| L1 | Level-1 |
| LDME | Long-Distance Matrix Element |
| LHC | Large Hadron Collider |
| LHCb | LHC beauty |
| LIP | Laboratório de Instrumentação e Física Experimental de Partículas |
| LO | Leading Order |
| LSB | Left SideBand |
| MB | Minimum Bias |
| MC | Monte Carlo |
| MCMC | Markov Chain Monte Carlo |
| MH | Metropolis–Hastings |
| ML | Maximum Likelihood |
| MPV | Most Probable Value |
| NLO | Next-to-Leading Order |
| NNLO | Next-to-Next-to-Leading Order |
| NP | Non-Prompt |
| NPLSB | Non-Prompt Left SideBand |
| NPRSB | Non-Prompt Right SideBand |
| NPSR | Non-Prompt Signal Region |
| NRQCD | Non-Relativistic Quantum ChromoDynamics |
| NUP | NUisance Parameter |
| ÖAW | Österreichische Akademie der Wissenschaften/Austrian Academy of Sciences |
| PAG | Physics Analysis Group |
| PAS | Physics Analysis Summary |
| PD | Primary Dataset |
| PDF | Probability Density Function |
| PKU | Beijing University |
| POI | Parameter Of Interest |
| $pp$ | proton–proton |
| $p\bar{p}$ | proton–anti-proton |
| PPD | Posterior Probability Density |
| PR | PRompt |

| | |
|---|---|
| PRLSB | PRompt Left SideBand |
| PRRSB | PRompt Right SideBand |
| PRSR | PRompt Signal Region |
| PU | Pile-Up |
| PV | Primary Vertex |
| PX | Perpendicular Helicity |
| QCD | Quantum ChromoDynamics |
| QED | Quantum ElectroDynamics |
| QGP | Quark Gluon Plasma |
| RHIC | Relativistic Heavy Ion Collider |
| RMS | Root Mean Square |
| RPC | Resistive Plate Chamber |
| RSB | Right SideBand |
| SB | SideBand |
| SDC | Short-Distance Coefficient |
| SM | Standard Model |
| SP | Single Photon |
| SR | Signal Region |
| SV | Secondary Vertex |
| T0 | Tier-0 |
| T1 | Tier-1 |
| T2 | Tier-2 |
| T&P | Tag and Probe |
| TEC | Tracker EndCap |
| TIB | Tracker Inner Barrel |
| TID | Tracker Inner Disk |
| TOB | Tracker Outer Barrel |

# Chapter 1
# Introduction

The goal of the scientific discipline of particle physics is to understand the principles that govern the universe at its most fundamental level. The standard model of particle physics (SM)[1] describes the current understanding of the elementary particles and the interactions among them, and represents the most accurate description of nature that we know today.

There are four known fundamental forces in nature, the electromagnetic, gravitational, weak and strong interactions. Each can be interpreted as resulting from the dynamics of a physical field, while the elementary particles can be understood as excited states of the corresponding fields. The fundamental (anti)matter particles are fermions, and consist of the (anti)leptons and (anti)quarks, while the interactions between the (anti)matter particles are mediated by the fundamental gauge bosons, so-called force carriers. The SM incorporates the particles and their interactions through the theoretical framework of the quantum field theory. However, there is no experimentally verified quantum theory of gravity, which is represented by a classical field within the general theory of relativity [2], and is not part of the SM.

The basis for the SM was set with the gradual transformation of the classical understanding of electrodynamics into a relativistic quantum field theory named "quantum electrodynamics" (QED) in the 1940s, which was made possible after combining the revolutionary ideas of quantum mechanics and the special theory of relativity [3]. In the QED formulation, the massless spin-1 photon mediates the electromagnetic interaction between electrically charged particles.

Given the success of QED, efforts were started to understand the weak interaction, responsible for radioactive decays, by a similar, QED-inspired field theory. In the 1960s, these attempts culminated in a unified field theory for the electromagnetic and the weak interactions [4–6], which postulated three further spin-1 bosons, mediating

---

[1] The SM is described in detail in many textbooks, see for example Ref. [1].

© Springer International Publishing Switzerland 2017

V. Knünz, *Measurement of Quarkonium Polarization to Probe QCD at the LHC*, Springer Theses, DOI 10.1007/978-3-319-49935-2_1

1

the weak interaction, the $Z^0$, $W^+$ and $W^-$ bosons, which were eventually discovered at CERN [7–10]. Due to the short-range nature of the weak interaction, its force carriers are required to be massive, which posed a serious problem to the theoretical framework. The "Brout–Englert–Higgs mechanism" [11–13] was proposed as a solution to this conundrum, predicting the spin-0 Higgs boson that was eventually discovered at CERN [14, 15] almost half a century later, confirming the proposed mechanism to be realized in nature.

Finally, the understanding of the strong interaction was accelerated by the introduction of the concept of quarks as the fundamental constituents of hadrons [16, 17]. Following this, in the 1970s a quantum field theory describing the strong interaction was formulated, "quantum chromodynamics" (QCD) [18]. In this model, the strong interaction is mediated by massless spin-1 gluons between the quarks, which carry a "color charge".

This completed the development of the SM, which can be regarded as a joint effort of theoreticians and experimentalists, as the main ideas were driven by both new theoretical developments as well as experimental discoveries. The description of particles and their interactions through the SM has been experimentally verified by many experiments over several decades with fantastic precision. Several particles and their properties have been predicted by the SM before they were discovered, giving further credence to the model.

Despite its impressive successes, the SM is not regarded as complete, and is thought to represent an intermediate step towards a more fundamental theory. The lack of reconcilability of the SM with general relativity is certainly one of its most profound shortcomings. Moreover, the SM has neither an explanation for dark matter, which contributes around 26% to the energy density of the observable universe [19], nor for the mysterious dark energy, contributing around 69% [19], which is thought to be responsible for the accelerated expansion of the universe. Furthermore, the SM has "aesthetic" problems: there is a large number of numerical constants which are not constrained by the SM itself. Also, the SM suffers from the so-called hierarchy problem, which emerges from the fact that gravity appears weaker than the other three interactions by many orders of magnitude, which leads to the necessity of fine-tuning of the model's parameters, challenging its "naturalness". Hence, huge past and ongoing efforts target the formulation and experimental verification of a more fundamental theory at higher energy scales that includes the SM in its low-energy limit.

Yet, the SM itself exhibits several problems that still remain to be solved. While the QED predictions are well understood and confirmed in an immensely precise manner, the strong interaction still poses many mysteries, which are yet to be understood. The research described in this document aims to contribute to this effort.

QCD describes the dynamics of the strong interaction in a self-consistent and compact way. However, concrete calculations can only be successfully performed in specific conditions. The level of the understanding of QCD processes is inevitably connected to the magnitude of the momentum transfer of the interacting particles (quarks and gluons), given the "running" strong coupling constant $\alpha_s$ [20, 21], limiting the possibility to perform perturbative calculations as an expansion in orders of $\alpha_s$ only for processes involving high momentum transfers at short distances ("hard scattering"), where $\alpha_s$ turns out to be small. At the other end of the energy scale, in the realm of "soft QCD", characterized by processes involving small momentum transfers at long distances, the expansion breaks down, and perturbative calculations are no longer possible, limiting the predictive power of QCD in these conditions.

Processes leading to the formation of QCD bound states ("hadronization") are difficult to access with QCD calculations, given that these processes are characterized by low momentum transfers [22]. This issue is mitigated in cases where only heavy quarks Q are involved (charm or beauty quarks, denoted as $c$ and $b$, respectively), as is the case for quarkonium states, which are QCD bound states of a heavy quark and its respective antiquark. Due to the heavy masses of the $c$ and $b$ quarks, the production of these quarkonium states is conjectured to be well factorized in two steps, occurring at two distinct time scales. The short-distance parton-level strong interaction processes, calculable within perturbative QCD and responsible for the production of an initial quark-antiquark pair $Q\bar{Q}$, are followed by the well separated hadronization of this $Q\bar{Q}$ pair into the QCD bound quarkonium state, where the $Q\bar{Q}$ pair undergoes long-distance strong interactions, part of the non-perturbative regime of QCD. Even though non-perturbative effects are important for the description of quarkonium production, the clear separation of the short-distance and long-distance effects allows the calculation of measurable observables, within the so-called non-relativistic QCD (NRQCD) factorization approach [23].

NRQCD is a QCD-inspired effective field theory, fully relying on the factorization between short-distance and long-distance effects. The NRQCD model describes quarkonium production as a superposition of the production of various "pre-resonant" $Q\bar{Q}$ states, characterized by different spin, angular momentum and color eigenstates, including color-singlet (CS) and color-octet (CO) configurations. The long-distance effects, containing all non-perturbative physics involved, can be described by the so-called long-distance matrix elements (LDMEs), supposedly universal and constant parameters, determining the relative importance of the individual $Q\bar{Q}$ states with respect to the full quarkonium production cross section. The magnitude of the LDMEs can be estimated by NRQCD velocity scaling rules. The LDMEs of the individual intermediate $Q\bar{Q}$ states are proportional to certain powers of the relative quark velocity $v$, which is rather small, given the heavy quark masses. Therefore, the non-perturbative hadronization process can be considered as an expansion in powers of $v$, limiting the number of contributing intermediate $Q\bar{Q}$ states, assuming a non-relativistic approximation [23]. The predictive power of NRQCD is limited given that the LDMEs are not calculable, and have to be determined from fits to experimental data. Several experimentally accessible physics observables are sensitive to the numerical values of the LDMEs and can therefore be used to perform their determinations by comparing theory calculations and data.

The most important of those quantities are the quarkonium production cross sections, differential in transverse momentum $p_T$, and the quarkonium polarizations, revealing information about the preferred spin alignment. The Tevatron experiments measured quarkonium cross sections [24–26] and polarizations [27–30]. The cross section measurements contributed vastly to the understanding of quarkonium production, given that they were precise enough to show that intermediate color-octet states significantly contribute to quarkonium production [31]. Before these measurements were available, it was conceived as likely that the color-singlet intermediate $Q\bar{Q}$ state was the only contributing source. However, the Tevatron quarkonium polarization measurements caused considerable troubles in the field, also known as the "quarkonium polarization puzzle", an unfortunate situation caused by two issues. Firstly, the Tevatron experiments did not observe any strong polarizations, seriously challenging the NRQCD factorization approach, which predicted almost fully transverse polarization [32]. Secondly, the Tevatron experiments published measurements of quarkonium polarization that were mutually inconsistent, and can therefore not be reliably interpreted. In fact, the analysis strategies of these Tevatron measurements were, a posteriori, identified as being ambiguous, by a series of papers [33–38] that introduced a new quarkonium polarization formalism and led to improved methodologies for reliable, robust and unambiguous measurements of quarkonium polarization.

Considering this inauspicious situation of quarkonium production physics in the Tevatron era, the ultimate objective in this field of research is a full clarification of the two above mentioned issues, requiring both experimental and phenomenological efforts. The challenging mission to clarify these open issues is therefore twofold. Firstly, the quarkonium polarization observables need to be understood experimentally in an unambiguous way, extending the measurements to the highest possible transverse momentum. Secondly, in case the observed discrepancies of the measured polarization parameters with NRQCD calculations persist, these discrepancies need to be understood from a phenomenological point of view, aiming at a coherent picture of quarkonium production cross sections and polarizations.

The Large Hadron Collider (LHC) provides ideal conditions to finally establish a clear experimental picture of quarkonium polarization, given the large quarkonium production rates, caused by the high center-of-mass (CM) energies $\sqrt{s}$ and proton-proton ($pp$) collision rates. Moreover, the Compact Muon Solenoid (CMS) detector is ideally suited to perform measurements of quarkonium cross sections and polarizations, up to very high $p_T$, due to the excellent muon momentum and vertex resolutions, and its efficient and flexible trigger system. The first core topic of this thesis is the detailed description of the measurement of the polarizations of all S-wave quarkonium states in $pp$ collisions at $\sqrt{s} = 7$ TeV with the CMS detector [39, 40], in the experimentally very clean decay of S-wave quarkonia into two muons. Following the recipes provided for a reliable and unambiguous measurement of quarkonium polarization, the decay angular distributions are analyzed in their full 2-dimensional form, additionally providing information about frame-invariant parameters. With respect to the Tevatron measurements, the CMS analyses can be performed with better precision and up to higher values of $p_T$, providing crucial information towards a deep understanding of quarkonium production.

Besides this experimental progress in the understanding of quarkonium polarization, NRQCD calculations were extended from leading order (LO) in $\alpha_s$ to next-to-leading order (NLO) [41–47], providing a more rigorous framework for the tests of NRQCD calculations. Moreover, the experimental results of the analyses described in this thesis, together with other results in the quarkonium production sector from the

LHC experiments, can be interpreted in a much more rigorous way than previously possible, due to the higher $p_T$ reach and the usage of more reliable experimental techniques for the estimation of the polarizations of the quarkonium states. Several state-of-the-art NLO NRQCD analyses can be found in the literature, affected by various problems leading to contradictory results for the estimated LDMEs [47–53]. The second core topic of this thesis is an attempt to clarify this situation by introducing an original phenomenological interpretation of the LHC results [54], characterized by a meticulously developed global fit framework, providing the possibility for a detailed study of the quarkonium production cross section and polarization observables, simultaneously and consistently treating these highly correlated measurements.

The thesis is organized as follows. The basics of quarkonium production physics, including the formal introduction of the NRQCD factorization approach, as well as the experimental situation of quarkonium production physics in the pre-LHC era, are discussed in Chap. 2, motivating the need for further measurements of quarkonium production observables at the LHC experiments, as well as phenomenological clarifications. The experimental setup, including the LHC accelerator and the CMS detector, are introduced in Chap. 3, describing in detail the CMS sub-detector systems, the trigger and the reconstruction strategies relevant for this thesis. Chapter 4 provides a description of the measurements of the $\Upsilon(nS)$ and $\psi(nS)$ polarizations with the CMS experiment, discussing in detail the general analysis strategy, the systematic uncertainties and the corresponding results. These results are then discussed and interpreted in Chap. 5, including a review of the LHC quarkonium production data, a review of the existing NRQCD analyses attempting to interpret these results and, most importantly, a detailed description of the original phenomenological interpretation of the LHC quarkonium production results as a product of this work.

# References

1. Griffiths DJ (2008) Introduction to elementary particles; 2nd rev. version. Physics textbook. Wiley, New York. ISBN 3-527-40601-8
2. Einstein A (1916) Die Grundlage der allgemeinen Relativitätstheorie. Ann der Phys 49:769
3. Einstein A (1905) Zur Elektrodynamik bewegter Körper. Ann der Phys 17:891
4. Glashow SL (1961) Partial-symmetries of weak interactions. Nucl Phys 22:579
5. Weinberg S (1967) A model of leptons. Phys Rev Lett 19:1264
6. Salam A, Ward JC (1964) Electromagnetic and weak interactions. Phys Lett 13:168
7. UA1 Collaboration (1983) Experimental observation of isolated large transverse energy electrons with associated missing energy at $\sqrt{s} = 540$ GeV. Phys Lett B 122:103
8. UA1 Collaboration (1983) Experimental observation of lepton pairs of invariant mass around 95 GeV/$c^2$ at the CERN SPS collider. Phys Lett B 126:398
9. UA2 Collaboration (1983) Observation of single isolated electrons of high transverse momentum in events with missing transverse energy at the CERN pp collider. Phys Lett B 122:476
10. UA2 Collaboration (1983) Evidence for $Z^0 \rightarrow e^+e^-$ at the CERN pp collider. Phys Lett B 129:130
11. Higgs PW (1964) Broken symmetries and the masses of gauge bosons. Phys Rev Lett 13:508

12. Englert F, Brout R (1964) Broken symmetry and the mass of gauge vector mesons. Phys Rev
    Lett 13:321
13. Guralnik GS, Hagen CR, Kibble TWB (1964) Global conservation laws and massless particles.
    Phys Rev Lett 13:585
14. CMS Collaboration (2012) Observation of a new boson at a mass of 125 GeV with the CMS
    experiment at the LHC. Phys Lett B 716:30
15. ATLAS Collaboration (2012) Observation of a new particle in the search for the standard model
    higgs boson with the ATLAS detector at the LHC. Phys Lett B 716:1
16. Gell-Mann M (1964) A schematic model of baryons and mesons. Phys Lett 8:214
17. Zweig G (1964) An SU(3) model for strong interaction symmetry and its breaking. CERN-
    TH-412
18. Fritzsch H, Gell-Mann M, Leutwyler H (1973) Advantages of the color octet gluon picture.
    Phys Lett B 47:365
19. Planck Collaboration (2015) Planck 2015 results. I. Overview of products and scientific results
20. Gross DJ, Wilczek F (1973) Ultraviolet behavior of non-abelian gauge theories. Phys Rev Lett
    30:1343
21. Politzer HD (1973) Reliable perturbative results for strong interactions? Phys Rev Lett 30:1346
22. QWG Collaboration (2011) Heavy quarkonium: progress, puzzles, and opportunities. Eur Phys
    J C 71:1534
23. Bodwin GT, Braaten E, Lepage GP (1995) Rigorous QCD analysis of inclusive annihilation
    and production of heavy quarkonium. Phys Rev D 51:1125
24. CDF Collaboration (1996) Quarkonia production at CDF. Nucl Phys A 610:373C
25. CDF Collaboration (1997) $J/\psi$ and $\psi(2S)$ production in $p\bar{p}$ collisions at $\sqrt{s} = 1.8$ TeV. Phys
    Rev Lett 79:572
26. CDF Collaboration (1997) Production of $J/\psi$ Mesons from $\chi_c$ Meson Decays in $p\bar{p}$ collisions
    at $\sqrt{s} = 1.8$ TeV. Phys Rev Lett 79:578
27. CDF Collaboration (2000) Measurement of $J/\psi$ and $\psi(2S)$ polarization in $p\bar{p}$ collisions at
    $\sqrt{s} = 1.8$ TeV. Phys Rev Lett 85:2886
28. CDF Collaboration (2007) Polarization of $J/\psi$ and $\psi(2S)$ mesons produced in $p\bar{p}$ collisions
    at $\sqrt{s} = 1.96$ TeV. Phys Rev Lett 99:132001
29. CDF Collaboration. CDF Public Note 9966
30. D0 Collaboration (2008) Measurement of the polarization of the $\Upsilon(1S)$ and $\Upsilon(2S)$ states in
    $p\bar{p}$ collisions at $\sqrt{s} = 1.96$ TeV. Phys Rev Lett 101:182004
31. Krämer M (2001) Quarkonium production at high-energy colliders. Prog Part Nucl Phys 47:141
32. Braaten E, Kniehl B, Lee J (2000) Polarization of prompt $J/\psi$ at the Tevatron. Phys Rev D
    62:094005
33. Faccioli P, Lourenço C, Seixas J, Wöhri H (2009) J/$\psi$ polarization from fixed-target to collider
    energies. Phys Rev Lett 102:151802
34. Faccioli P, Lourenço C, Seixas J (2010) Rotation-invariant relations in vector meson decays
    into fermion pairs. Phys Rev Lett 105:061601
35. Faccioli P, Lourenço C, Seixas J (2010) A new approach to quarkonium polarization studies.
    Phys Rev D 81:111502
36. Faccioli P, Lourenço C, Seixas J, Wöhri H (2011) Model-independent constraints on the shape
    parameters of dilepton angular distributions. Phys Rev D 83:056008
37. Faccioli P, Lourenço C, Seixas J, Wöhri H (2010) Towards the experimental clarification of
    quarkonium polarization. Eur Phys J C 69:657
38. Faccioli P, Lourenço C, Seixas J, Wöhri H (2011) Determination of $\chi_c$ and $\chi_b$ polarizations
    from dilepton angular distributions in radiative decays. Phys Rev D 83:096001
39. CMS Collaboration (2013) Measurement of the $\Upsilon(1S)$, $\Upsilon(2S)$ and $\Upsilon(3S)$ polarizations in $pp$
    collisions at $\sqrt{s} = 7$ TeV. Phys Rev Lett 110:081802
40. CMS Collaboration (2013) Measurement of the prompt $J/\psi$ and $\psi(2S)$ polarizations in pp
    collisions at $\sqrt{s} = 7$ TeV. Phys Lett B 727:381
41. Artoisenet P, Lansberg JP, Maltoni F (2007) Hadroproduction of $J/\psi$ and $\Upsilon$ in association
    with a heavy-quark pair. Phys Lett B 653:60

42. Campbell JM, Maltoni F, Tramontano F (2007) QCD corrections to $J/\psi$ and Upsilon production at hadron colliders. Phys Rev Lett 98:252002
43. Gong B, Wang JX (2008) Next-to-leading-order QCD corrections to $J/\psi$ polarization at Tevatron and large-hadron-collider energies. Phys Rev Lett 100:232001
44. Gong B, Wang JX (2008) QCD corrections to polarization of $J/\psi$ and $\Upsilon$ at Tevatron and LHC. Phys Rev D 78:074011
45. Gong B, Li XQ, Wang JX (2009) QCD corrections to $J/\psi$ production via color octet states at Tevatron and LHC. Phys Lett B 673:197
46. Ma YQ, Wang K, Chao KT (2011) QCD radiative corrections to $\chi_{cJ}$ production at hadron colliders. Phys Rev D 83:111503
47. Butenschön M, Kniehl B (2012) $J/\psi$ polarization at Tevatron and LHC: nonrelativistic-QCD factorization at the crossroads. Phys Rev Lett 108:172002
48. Butenschön M, Kniehl B (2011) World data of $J/\psi$ production consolidate NRQCD factorization at NLO. Phys Rev D 84:051501
49. Gong B, Wan LP, Wang JX, Zhang HF (2013) Polarization for prompt $J/\psi$, $\psi(2S)$ production at the Tevatron and LHC. Phys Rev Lett 110:042002
50. Gong B, Wan LP, Wang JX, Zhang HF (2014) Complete next-to-leading-order study on the yield and polarization of $\Upsilon(1S, 2S, 3S)$ at the Tevatron and LHC. Phys Rev Lett 112:032001
51. Ma YQ, Wang K, Chao KT (2011) $J/\psi$ ($\psi'$) production at the Tevatron and LHC at $\mathcal{O}(\alpha_s^4 v^4)$ in nonrelativistic QCD. Phys Rev Lett 106:042002
52. Chao KT, Ma YQ, Shao HS, Wang K, Zhang YJ (2012) $J/\psi$ polarization at hadron colliders in nonrelativistic QCD. Phys Rev Lett 108:242004
53. Han H, Ma YQ, Meng C, Shao HS, Zhang YJ, Chao KT (2014) $\Upsilon(nS)$ and $\chi_b(nP)$ production at hadron colliders in nonrelativistic QCD. arXiv:1410.8537
54. Faccioli P, Knünz V, Lourenço C, Seixas J, Wöhri H (2014) Quarkonium production in the LHC cra: a polarized perspective. Phys Lett B 736:98

# Chapter 2
# Quarkonium Physics

This chapter is structured as follows. Section 2.1 introduces the quarkonium spectrum, its decays, and a summary of the fields of research that use quarkonia as probes for their physics objectives. A detailed discussion of the current state-of-the-art model calculations for quarkonium production is presented in Sect. 2.2, including a historical approach to the problem, as well as comparisons of these models with data from the pre-LHC era. Finally, Sect. 2.3 will discuss recent progress regarding the analysis methodologies to be employed in measurements of quarkonium polarization.

## 2.1 Introduction

Quarkonia are bound states of a heavy quark and its respective antiquark, $Q\bar{Q}$, bound by the strong force. These mesons appear in two distinct "families" of states, the charmonium system, containing the mesons consisting of two charm quarks, $c\bar{c}$, and the bottomonium system, containing the mesons consisting of two beauty quarks, $b\bar{b}$.

**Quarkonium Spectrum**

The $Q\bar{Q}$ bound system is realized in nature in many different quantum states, characterized by the quantum numbers describing the angular momentum $L$, the spin $S$, the total angular momentum $J = S + L$, and the principal quantum number $N$. Notations in the literature include both the $J^{PC}$ convention, with parity $P = (-1)^{(L+1)}$ and charge conjugation $C = (-1)^{(L+S)}$, as well as the spectroscopic notation $N^{2S+1}L_J$.

Figures 2.1 and 2.2 show a summary of the charmonium and bottomonium systems, respectively, showing a subset of the quarkonium states relevant for this thesis and a subset of the decays that occur within the families. Decays $b\bar{b} \rightarrow c\bar{c}$ can be neglected [1]. These figures are restricted to CP-even states, $J^{++}$ and $J^{--}$, below the open charm and open beauty thresholds. The CP-odd $0^{-+}$ ($\eta_c$ and $\eta_b$) and $1^{+-}$ ($h_c$ and $h_b$) states are not discussed in this thesis. The quarkonium spectra can be divided

© Springer International Publishing Switzerland 2017
V. Knünz, *Measurement of Quarkonium Polarization to Probe QCD at the LHC*, Springer Theses, DOI 10.1007/978-3-319-49935-2_2

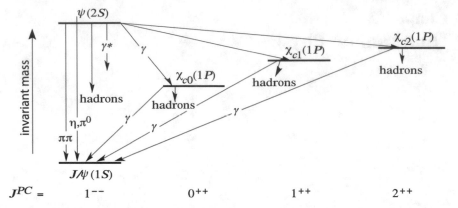

**Fig. 2.1** Charmonium spectrum and decays, adapted from Ref. [1], limited to the CP-even states below the open charm threshold

in two main categories, characterized by two different experimental signatures, the "S-wave" ($L = 0$) and "P-wave" ($L = 1$) states.[1] The S-wave states are the $J^{PC} = 1^{--}$ vector mesons J/$\psi$ and $\Upsilon(1S)$, and their radial excitations $\psi(2S)$[2] and the $\Upsilon(2S)$ and $\Upsilon(3S)$ mesons, respectively. The two charmonium S-wave states are referred to as $\psi(nS)$ (with $n = 1, 2$) states, while the three S-wave bottomonium states are denoted as $\Upsilon(nS)$ (with $n = 1, 2, 3$) states. The P-wave states are the $J^{PC} = J^{++}$ pseudo vector mesons $\chi_{cJ}$ and $\chi_{bJ}(1P)$ that appear in triplets corresponding to $J = 1, 2, 3$ and their radial excitations, in the case of the bottomonium system, the $\chi_{bJ}(2P)$ and $\chi_{bJ}(3P)$. Experimentally, the most important decay modes are the "dimuon" decays of the S-wave states, $\psi(nS) \rightarrow \mu\mu$ and $\Upsilon(nS) \rightarrow \mu\mu$, and the radiative decays of the P-wave states, $\chi_{cJ} \rightarrow$ J/$\psi + \gamma$ and $\chi_{bJ}(nP) \rightarrow \Upsilon(mS) + \gamma$.

Detailed listings of the particle masses, full widths, decay modes and the corresponding branching fractions can be found in Ref. [1]. For the charmonium system, the masses cover the range from 3.0969 GeV of the J/$\psi$ up to 3.6861 GeV for the $\psi(2S)$; in the bottomonium system the masses cover the range from 9.4603 GeV of the $\Upsilon(1S)$ up to 10.534 GeV[3] for the $\chi_b(3P)$. The full widths of the quarkonium states are small compared to experimental resolution, except for the $\chi_{cJ}$ states, with widths of 10.3, 0.86 and 1.97 MeV, respectively for the $\chi_{c0}$, $\chi_{c1}$ and $\chi_{c2}$ states, while the widths of the $\chi_{bJ}(nP)$ states are yet to be measured. The decay times of the S-wave states are in the range of $2–40 \times 10^{-21}$ s, while the decay times of the P-wave states (in the charmonium cases, where they are measured) are in the range $6–70 \times 10^{-23}$ s. With these decay times, the quarkonium states only travel average distances of the order of femto- up to pico-meters, before they decay. All the quarkonium decays are therefore classified as "prompt" (quasi-instantaneous, PR) decays.

---

[1] In this thesis, the terms "S-wave" and "P-wave" states only refer to CP-even states.

[2] The $\psi(2S)$ state is also referred to as $\psi'$ in the literature.

[3] This PDG [1] mass value will be updated with recent LHCb results [2, 3].

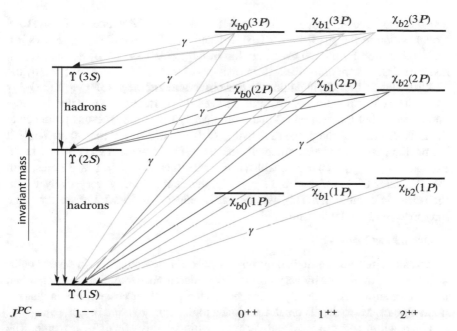

**Fig. 2.2** Bottomonium spectrum and decays, adapted from Ref. [1], limited to the CP-even states below the open beauty threshold. Decays of the type $\Upsilon(nS) \to \chi_{bJ}(mP) + \gamma$ are not shown

At this point it should be mentioned that natural units are used in this thesis, with $\hbar = c = 1$, therefore giving energy, mass and momentum in units of eV.

The experimental samples of S-wave and P-wave events are a mixture of directly produced quarkonia and products of the decays from heavier states, so-called "feed-down decays". While it is possible to separate, experimentally, samples of P-wave states, by requesting a $\gamma$ in the final state, it is not (currently) possible, due to the short decay times of quarkonia, to separate the directly produced S-wave states from the feed-down decays. The experimental measurements of S-wave states are therefore limited to the measurement of the properties of the prompt components, not removing the feed-down contributions. The same is true for the P-wave states, which are also affected by feed-down decays from radiative transitions from the S-wave states. However, by measuring the properties of the feed-down states, connected with knowledge about the "feed-down fractions", defining the mixture of the production channels of the prompt samples, the properties of the directly produced quarkonium states are accessible a posteriori, in the interpretation of the measurements.

The highest-mass charmonium and bottomonium states, the $\psi(2S)$ and $\chi_b(3P)$ states, are considered to be free of feed-down decays, hence the measurement of the respective prompt samples allows access to the directly produced quarkonium states. The corresponding measurements are therefore especially important, given that the comparison with model calculations is simplified thanks to the experimentally very clear information. The $\Upsilon(3S)$ state has been regarded as feed-down free for several decades, until recently, due to the discovery of the $\chi_b(3P)$ state by the ATLAS Collaboration in the $\chi_b(3P) \rightarrow \Upsilon(1S) + \gamma$ and $\chi_b(3P) \rightarrow \Upsilon(2S) + \gamma$ decays [4], the confirmation by the D0 Collaboration in the $\chi_b(3P) \rightarrow \Upsilon(1S) + \gamma$ decay [5], and the measurement of the feed-down fractions of the $\chi_b(3P) \rightarrow \Upsilon(nS) + \gamma$ by the LHCb Collaboration [2], including the first observation of the $\chi_b(3P) \rightarrow \Upsilon(3S) + \gamma$ decay. The corresponding estimate of the feed-down fraction is affected by large uncertainties, but the data reveal that more than $1/3$ of $\Upsilon(3S)$ mesons produced at the LHC originate through the radiative $\chi_b(3P) \rightarrow \Upsilon(3S) + \gamma$ transition. While it has not been established experimentally that the $\chi_b(3P)$ state is in fact the third radial excitation of the $\chi_{bJ}(1P)$ triplet, with a triplet substructure of $J = 1, 2, 3$ states, with the $J^{PC}$ states $0^{++}$, $1^{++}$ and $2^{++}$, $\chi_{bJ}(3P)$, this is regarded as very likely within the scientific community. This assumption is reflected in Fig. 2.2, therefore to be interpreted and used with care.

### Quarkonia as Probes

Quarkonium mesons are studied by various scientific communities, in several collision systems, motivated by very different considerations. Here, the most important aspects are summarized briefly, to emphasize the wealth of possibilities in quarkonium physics. More details on all mentioned topics can be found in Ref. [6].

Quarkonium spectroscopy and decays constitute active fields of research. The spectrum of "conventional" quarkonia, as discussed above, is rather well understood. With the exception of the $\chi_b(3P)$ discovery and subsequent studies regarding the nature (triplet-substructure) of this state, a more accurate measurement of its mass, and an assessment of the branching fractions of its decays, the chapter of conventional quarkonium spectroscopy can be regarded as closed. However, there is a wealth of studies ongoing, both at b-factory experiments and hadron collider experiments, in the field of so-called "exotic quarkonium" physics. In the last decade, several such exotic quarkonium states have been discovered, and their quantum states determined, often through decays involving the $\psi(nS)$ and $\Upsilon(nS)$ quarkonium states. The first and most famous of these states is the neutral $X(3872)$, first discovered by the Belle Collaboration [7] and confirmed by several other collaborations. The LHCb Collaboration has measured the quantum numbers of this state to be $J^{PC} = 1^{++}$, in the decay $X(3872) \rightarrow J/\psi\pi\pi$ [8]. The properties of the $X(3872)$, as measured by the individual experiments, do not fit the expectations of a simple charmonium state. The nature of this state is still unclear. The exciting possible explanations include a loosely-bound molecule of two mesons, as well as a tightly-bound diquark-diantiquark bound system, which would require the existence of two neutral and one charged partner state, which have not been established yet, experimentally [6]. A further very interesting state is the charged $Z(4430)^{\pm}$, discovered by the Belle Collaboration [9] and confirmed by the LHCb Collaboration [10]. The confirmation of this state has attracted attention of a wide range of the physics community, as the minimal quark content of such a charged state is $c\bar{c}d\bar{u}$ [10]. This can be be viewed as the first unambiguous evidence for hadrons with more than the traditional $q\bar{q}$ or $qqq/\bar{q}\bar{q}\bar{q}$ content, which was already proposed by Gell-Mann, in the original paper introducing quarks as the fundamental constituents of all hadrons [11].

Quarkonia also play a major role in experiments studying hot and dense QCD matter with heavy-ion (HI) collisions. In these collisions, at very high energy densities, a phase transition to a quark gluon plasma (QGP) is expected to occur. Quarkonia are produced very early in the collisions, prior to formation of the QGP. Their evolution through the medium produced in the HI collisions can provide information about the QGP. Due to a Debye screening of the QCD potential binding the $Q\bar{Q}$ pairs, quarkonia are expected to be "melted" in the hot medium [12]. Given that the individual quarkonia have very different binding energies, increasing with the difference of the quarkonium mass with respect to the open charm/beauty thresholds, the individual quarkonium states melt at different energy densities of the collisions. The higher-mass states get suppressed at lower energy densities than the lower-mass states, which are more tightly-bound. Therefore, one expects a sequential suppression of the quarkonium states as a function of the energy density [13], a smoking gun signal for QGP, affecting also the lower-mass states due to the suppression of the feed-down contributions. However, at the LHC other effects complicate the interpretation of the results, such as recombination [14], where due to the high abundance of charm quarks in the collisions, $c$ and $\bar{c}$ quarks produced in different nucleon-nucleon collisions bind together forming a charmonium state.

There are several other fields that use quarkonia as probes, which will not be discussed in more detail here. These topics include for example the measurement of CP violating phases in B-hadron decays, which often involve $\psi(nS)$ mesons in their final state signatures. Other interesting decays, which are however by far not yet accessible with the data samples collected by the LHC experiments, are the Higgs decays $H \rightarrow J/\psi + \gamma$ and $H \rightarrow \Upsilon(1S) + \gamma$, from which the $Hc\bar{c}$ and $Hb\bar{b}$ couplings can be measured [15].

Finally, motivating the research presented in this thesis, a detailed understanding of the fundamental mechanisms that lead to the production of quarkonia helps to understand hadron formation in general, which is not yet well understood in the SM, and is therefore an active field of research in both experiment and theory. The corresponding strategies for the model calculations of quarkonium production observables, including quarkonium cross sections and polarizations, are discussed in detail below.

## 2.2 Quarkonium Production

Due to their simple and symmetric composition, as well as to the heavy quark masses $m_Q$, heavy quarkonium states are ideal laboratories to test the interplay between perturbative and non-perturbative QCD. A detailed understanding of quarkonium production helps to understand hadron formation, how the strong interaction binds quarks into hadrons.

One basic concept guides all considerations regarding the understanding of quarkonium production. The production of any quarkonium state is assumed to be factorizable in two parts. The first part is the production of an intermediate $Q\bar{Q}$ pair at short distance, which is calculable within perturbative QCD and is fairly well understood. The second part is the hadronization, the intermediate $Q\bar{Q}$ pair forming a QCD bound quarkonium state. This stage of the production is part of the non-perturbative realm of QCD, which causes problems in the modeling and the understanding of quarkonium production.

Quarkonia can be treated as approximately non-relativistic systems, given the heavy quark masses and the resulting relative quark velocity $v$ in the bound state, with $v^2 \approx 0.3$ for the charmonium and $v^2 \approx 0.1$ for the bottomonium states [6]. Due to the relatively small heavy-quark velocities, the two factorized steps occur at distinct timescales. The time needed for the production of the $Q\bar{Q}$ pair is proportional to $1/m_Q$, while the non-perturbative formation of the bound state occurs at a time scale of the order of $1/(m_Q v^2)$ [16]. If these two timescales are well separated, which is the case if $1/(m_Q v^2) \gg 1/m_Q$, the intuitive expectation is that the short-distance and long-distance effects can indeed be separated. While this condition is well fulfilled for bottomonium states, and reasonably well for charmonium states, this is not the case, for example, for light hadrons, where the two production steps cannot be factorized in two distinct processes that occur at different time scales. For these reasons, quarkonia provide a unique opportunity to study hadron formation, and to learn about the interaction dynamics involving the long-distance strong force.

Full QCD calculations of quarkonium production observables are limited to the perturbative part of quarkonium production, up to certain powers in $\alpha_s$. The non-perturbative formation of the bound state is not calculable with perturbative approaches. This part would, in principle, be accessible by calculations within the framework of lattice QCD, but such an effort has not yet been performed. Therefore, current calculations for quarkonium production have to rely on certain assumptions and approximations. There are several different models, attempting to calculate quarkonium production observables, with various levels of success of reconciling data and model calculations. There is wide consensus in the scientific community that the NRQCD factorization approach currently provides the most reliable calculations, with the best chances of successfully describing simultaneously all available quarkonium production measurements. Therefore, this model is introduced in detail below, while other models, that would nevertheless deserve the attention of the reader, are not discussed here, but are summarized in Ref. [6]. The Color-Singlet Model (CSM) is included in the NRQCD factorization approach as a special case. The alternative approaches, not discussed here, include the Color-Evaporation Model (CEM), introduced in [17], and the $k_T$ factorization approach [6].

## 2.2.1   Non-relativistic QCD Factorization

NRQCD is an effective field theory that was introduced in Ref. [16]. NRQCD, in general, can be regarded as a direct consequence of full QCD, in the limit of $m_Q \rightarrow \infty$. However, the approach relies heavily on the validity of the factorization of the production of the initial $Q\bar{Q}$ pair, and the formation of the bound state. In the framework

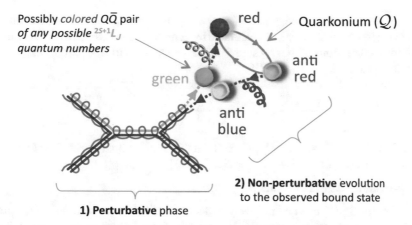

Possibly *colored* $Q\bar{Q}$ pair
*of any possible* $^{2S+1}L_J$
*quantum numbers*

red

Quarkonium ($\mathcal{Q}$)

green

anti
red

anti
blue

**2) Non-perturbative** evolution
to the observed bound state

**1) Perturbative** phase

**Fig. 2.3** Sketch illustrating the two distinct steps of quarkonium production: the perturbative production of an initial possibly colored $Q\bar{Q}$ pair, followed by the formation of a color-neutral quarkonium state $\mathcal{Q}$ via the non-perturbative emission of soft gluons [18]

of NRQCD, the cross section $\sigma(\mathcal{Q})$ of the quarkonium $\mathcal{Q}$ can be calculated by the simple factorization formula [16]

$$\sigma(\mathcal{Q}) = \sum_n \mathcal{S}[Q\bar{Q}(n)] \cdot \mathcal{O}^{\mathcal{Q}}(n) . \tag{2.1}$$

The calculation is factorized in the short distance coefficients (SDC) $\mathcal{S}[Q\bar{Q}(n)]$, describing the perturbative production of the initial $Q\bar{Q}$ pair in quantum state $n = {}^{2S+1}L_J^{[C]}$, with $C$ the color multiplicity, and the long-distance matrix elements $\mathcal{O}^{\mathcal{Q}}(n)$, describing the non-perturbative evolution into the bound quarkonium $\mathcal{Q}$ in state $n'$, $\mathcal{Q}^{n'}$. The individual terms of this sum, characterized by the various intermediate states $n$ are denoted as the "partial cross sections" $\sigma(n) = \mathcal{S}[Q\bar{Q}(n)] \cdot \mathcal{O}^{\mathcal{Q}}(n)$. The sum of the partial cross sections runs over all possible intermediate $Q\bar{Q}$ states $n$, including color-singlet (CS, $C = 1$) and color-octet (CO, $C = 8$) configurations. This formalism allows the existence of intermediate CO states in nature, with transitions into the physical color-neutral quarkonium bound state via the non-perturbative emission of soft gluons, as is illustrated in Fig. 2.3.

**Short-Distance Coefficients**

The SDCs $\mathcal{S}[Q\bar{Q}(n)]$ can be calculated with perturbative QCD approaches, as expansion in powers of $\alpha_s$, and correspond to the sum of the partonic cross sections to produce a $Q\bar{Q}$ pair in state $n$, convoluted with the parton distribution functions. The SDCs are functions accounting for the kinematic dependence of the cross section and decay distributions and are process dependent, different for any collision system and CM energy. As this thesis discusses measurements at the LHC, considerations are restricted to SDCs calculated for $pp$ collisions.

The polarization parameters $\vec{\lambda} = (\lambda_\vartheta, \lambda_\varphi, \lambda_{\vartheta\varphi})$, with respect to a certain quantization axis $z$ (see Sect. 2.3), can be calculated by defining the SDCs corresponding to the individual projections of the $Q\bar{Q}$ spin $S$ on the quantization axis $z$, $\mathcal{S}_{ij}[Q\bar{Q}(n)](= \mathcal{S}_{ij})$, with the notation $i, j = 0, \pm 1$ [19],

$$\lambda_\vartheta = \frac{\mathcal{S}_{11} - \mathcal{S}_{00}}{\mathcal{S}_{11} + \mathcal{S}_{00}}, \quad \lambda_\varphi = \frac{\mathcal{S}_{1,-1}}{\mathcal{S}_{11} + \mathcal{S}_{00}}, \quad \lambda_{\vartheta\varphi} = \frac{\sqrt{2}\mathrm{Re}\mathcal{S}_{10}}{\mathcal{S}_{11} + \mathcal{S}_{00}}, \tag{2.2}$$

with $\mathcal{S}_{11}$ being the "transverse SDC" and $\mathcal{S}_{00}$ the "longitudinal SDC". The total SDC is given by the relation $\mathcal{S}[Q\bar{Q}(n)] = \mathcal{S}_{00}[Q\bar{Q}(n)] + 2 \cdot \mathcal{S}_{11}[Q\bar{Q}(n)]$.

**Long-Distance Matrix Elements**

The LDMEs $\mathcal{O}^Q(n)$ can be intuitively understood as being proportional to the probability of a given intermediate $Q\bar{Q}$ in state $n$ to form a quarkonium state $Q$. They are constants, independent of the $Q\bar{Q}$ kinematics, and are assumed to be universal, identical for any collision system, only depending on the initial $Q\bar{Q}$ state $n$ and the final state $Q$. The LDMEs are not calculable with currently available techniques (with the exception of the CS LDMEs, see below), and have to be estimated by fits to experimental data, discussed in more detail in Sects. 2.2.2 and 5.2.

In principle, one would have to sum over all possible intermediate states $n$ in order to calculate the full, "color-inclusive" cross sections and polarizations of the individual quarkonium states $Q$. However, the individual LDMEs can be organized in certain hierarchies, "power-counting schemes" or "velocity scaling rules", which estimate the relative size of the individual LDMEs in powers of the heavy-quark velocity $v$. There are various slightly differing suggestions for these hierarchies in the literature. A fairly common definition of the importance of the individual LDMEs follows the relation [20]

$$\mathcal{O}^Q(n) \propto v^{2L+2E_1+4M_1}, \tag{2.3}$$

with $L$ the angular momentum of the $Q\bar{Q}$ state, $E_1$ the minimum number of chromoelectric ($\Delta L = \pm 1$, $\Delta S = 0$) transitions necessary to reach the quarkonium state $Q$ from the $Q\bar{Q}$ state $n$, and $M_1$ the minimum number of chromomagnetic ($\Delta L = 0$, $\Delta S = \pm 1$) transitions. Table 2.1 summarizes the most important states $n$ and the corresponding expected suppression, in powers of $v$, following these velocity scaling rules, separately for S-wave $1^{--}$ and P-wave $J^{++}$ states. This table includes CO intermediate states, as well as CS transitions with $S$, $L$ and $J$ configurations different from those of the final state $Q$, which are clearly suppressed and not considered in the CSM.

Due to the small velocities $v$ in the heavy quarkonium states, the partial cross sections of the states characterized by large powers of $v$ are expected to be negligible with respect to the leading ones. It has to be emphasized that even though the LDMEs of velocity scaling suppressed states $n$ are expected to be small, this could in principle be compensated by large values of the corresponding SDCs. However, the standard approach is to only consider in the sum over intermediate $Q\bar{Q}$ states

**Table 2.1** Expected scaling of the LDMEs $\mathcal{O}^{\mathcal{Q}}(n)$, in powers of $v$, for $1^{--}$ and $J^{++}$ states [20]

| CS states $n$ | $^1S_0^{[1]}$ | $^3S_1^{[1]}$ | $^1P_1^{[1]}$ | $^3P_J^{[1]}$ | $^3D_J^{[1]}$ | $^1D_2^{[1]}$ |
|---|---|---|---|---|---|---|
| $\mathcal{Q}=1^{--}$ | $v^8$ | $\mathbf{1}$ | $v^8$ | $v^8$ | $v^8$ | $v^{12}$ |
| $\mathcal{Q}=J^{++}$ | $v^6$ | $v^6$ | $v^{10}$ | $\mathbf{v^2}$ | $v^{10}$ | $v^{10}$ |

| CO states $n$ | $^1S_0^{[8]}$ | $^3S_1^{[8]}$ | $^1P_1^{[8]}$ | $^3P_J^{[8]}$ | $^3D_J^{[8]}$ | $^1D_2^{[8]}$ |
|---|---|---|---|---|---|---|
| $\mathcal{Q}=1^{--}$ | $\mathbf{v^4}$ | $\mathbf{v^4}$ | $v^8$ | $\mathbf{v^4}$ | $v^8$ | $v^{12}$ |
| $\mathcal{Q}=J^{++}$ | $v^6$ | $\mathbf{v^2}$ | $v^6$ | $v^6$ | $v^6$ | $v^{10}$ |

$n$ (Eq. 2.1) the states whose expected velocity scaling goes up to and including $v^4$. This approach is well justified for bottomonium states, due to the heavy mass of the beauty quark. However, it remains to be seen if these scaling rules are applicable also for charmonium states, with a considerably larger $v$. These considerations lead to the commonly considered intermediate states $^3S_1^{[1]}$, $^1S_0^{[8]}$, $^3S_1^{[8]}$ and $^3P_J^{[8]}$ for S-wave $1^{--}$ quarkonia, and the intermediate states $^3P_J^{[1]}$ and $^3S_1^{[8]}$ for P-wave $J^{++}$ quarkonia. The actually considered terms differ in the individual NRQCD analyses, as detailed in Sect. 5.2.

**Color-Singlet Model as Special Case of the NRQCD Factorization Approach**

The CSM can be obtained as a special case of the NRQCD factorization approach, if in Eq. 2.1 only the CS term is considered, which is characterized by identical states $n$ and $\mathcal{Q}$. In this case, for S-wave $1^{--}$ quarkonia only the $^3S_1^{[1]}$ intermediate state is considered, while for P-wave $J^{++}$ quarkonia only the $^3P_J^{[1]}$ state is considered. Considering the CSM alone leads to infrared divergencies in the case of the calculation of the cross sections of $J^{++}$ quarkonia, which can only be compensated by the addition of CO terms [16].

Due to the simplicity of the transition of the CS $Q\bar{Q}$ state into the quarkonium state $\mathcal{Q}$ ($\Delta L = 0$, $\Delta S = 0$, no non-perturbative emission of soft gluons), the CS LDMEs can be calculated with high precision in several ways, including potential model approaches [6], and determined experimentally, through the measurement of the decay widths of the quarkonium states, given the known relations between the production and decay matrix elements [21].

**Status of NRQCD Calculations**

Full NRQCD calculations exist at LO in $\alpha_s$ and NLO, for various quarkonium states, CS and CO channels, collision systems, kinematic regions and center-of-mass energies. The ones relevant for this work can be found in Refs. [19, 22–27]. Figure 2.4 shows dominant LO diagrams for the hadroproduction of $1^{--}$ quarkonia for the $^3S_1^{[1]}$ CS channel (left), the dominating gluon fragmentation diagram for the $^3S_1^{[8]}$ CO channel (middle), and a LO diagram for the $^1S_0^{[8]}$ and $^3P_J^{[8]}$ CO channels (right).

These perturbative QCD calculations are provided as a function of the quarkonium kinematics in the laboratory frame. In collider experiments it is common to use the

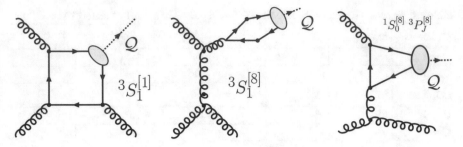

**Fig. 2.4** LO diagrams for the hadroproduction of $1^{--}$ quarkonia $\mathcal{Q}$ for the $^3S_1^{[1]}$ CS channel (*left*), for the $^3S_1^{[8]}$ CO channel (*middle*), and for the $^1S_0^{[8]}$ and $^3P_J^{[8]}$ CO channels (*right*) [6]

transverse momentum $p_T$ and the rapidity $y$, defined as

$$y = \frac{1}{2} \ln \left( \frac{E + p_L}{E - p_L} \right) , \tag{2.4}$$

with $E$ the particle energy and $p_L$ the longitudinal momentum, along the beam-axis. Regions in phase space close to $|y| = 0$ are referred to as "mid-rapidity" regions, contrary to "forward-rapidity" regions, characterized by larger values of $|y|$.

Figure 2.5 shows an example of such NRQCD calculations, as taken from Ref. [19], for $J/\psi$ production in $pp$ collisions at $\sqrt{s} = 7$ TeV at mid-rapidity, as a function of $p_T$. The top left panel shows the SDCs at NLO. The top right panel shows the ratio of the SDCs at NLO with respect to the LO calculations, indicating the relative change between the LO and NLO calculations. This ratio is denoted as the SDC "k-factor". The bottom panel shows the polarization parameter $\lambda_\vartheta$ in the Helicity (HX) frame at NLO. The polarization parameters and reference frames are defined in Sect. 2.3. These calculations are made for $|y| < 0.9$, but the calculations [19] have shown that the SDCs and polarizations of the individual color channels change only marginally with rapidity.

At NLO, the individual color channels have very different polarizations, ranging from almost fully longitudinal ($^3S_1^{[1]}$), unpolarized ($^1S_0^{[8]}$), almost fully transverse $^3S_1^{[8]}$ to "hyper-transverse" ($^3P_J^{[8]}$), with a divergent behavior and a change of sign at $p_T \approx 9$ GeV. Above a certain $p_T$ of approximately 15 GeV the polarizations at NLO can be regarded as constant. The shapes of the $p_T$-differential SDCs at NLO are rather similar for the individual color channels, albeit characterized by different levels of steepness of the curves as a function of $p_T$, the $^3S_1^{[1]}$ being the steepest, and the $^3S_1^{[8]}$ and $^3P_J^{[8]}$ being the flattest, towards high $p_T$. The NLO SDC of the $^3P_J^{[8]}$ is positive at low $p_T$ and changes sign at around 7 GeV (not visible in the top left panel of Fig. 2.5 due to the shown range in $p_T$).

Comparing the behavior at LO and NLO one can observe that the $^3S_1^{[8]}$ and $^1S_0^{[8]}$ are very stable in the perturbative expansion, with small SDC k-factors and no changes in polarization [19]. On the contrary, the $^3S_1^{[1]}$ and $^3P_J^{[8]}$ channels show very large

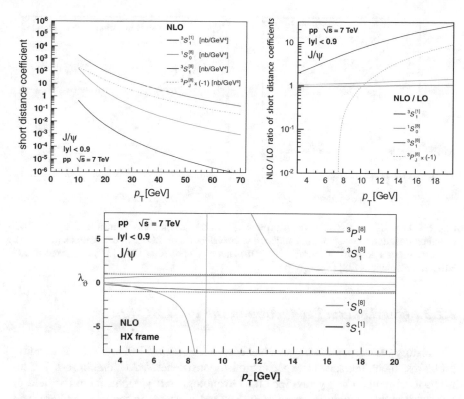

**Fig. 2.5** NRQCD calculations [19] for $J/\psi$ production in $pp$ collisions at $\sqrt{s} = 7$ TeV at mid-rapidity, as a function of $p_T$, showing the SDCs at NLO (*top left*), the ratio of the SDCs at NLO with respect to the LO calculations (*top right*) and the polarization parameter $\lambda_\vartheta^{HX}$ at NLO (*bottom*)

changes, indicating that the perturbative expansion in $\alpha_s$ is not yet convergent at NLO. Furthermore, the polarization parameter $\lambda_\vartheta$ changes from almost fully transverse at LO to almost fully longitudinal at NLO for the $^3S_1^{[1]}$ channel. For the $^3P_J^{[8]}$ channel, $\lambda_\vartheta$ changes from the unpolarized scenario at LO to a hyper-transverse polarization at NLO [19]. Given the large SDC k-factors of the $^3S_1^{[1]}$ and $^3P_J^{[8]}$ channels, it would be desirable to have access to calculations of higher-order QCD corrections, at next-to-next-to-leading order (NNLO), or even beyond. However, full NNLO calculations are beyond the scope of the techniques currently used. For the CS $^3S_1^{[1]}$ channel there has been a large effort to calculate partial NNLO corrections, denoted as NNLO* [28, 29]. These calculations take into account processes where the $1^{--}$ quarkonium is produced in association with three light partons, which are assumed, by the authors, to be the dominantly contributing processes at NNLO. The success of describing hadron collider quarkonium production data with this model, among others, is discussed below.

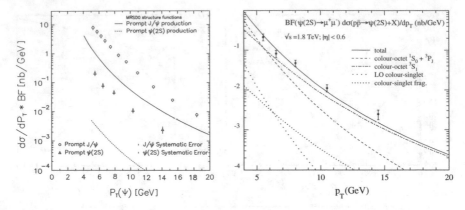

**Fig. 2.6** Production cross sections of prompt J/$\psi$ and $\psi(2S)$ mesons as measured by the CDF Collaboration, as a function of $p_T$, compared to curves based on LO CSM calculations [37] (*left*). Production cross sections of prompt $\psi(2S)$ mesons as measured at CDF [37], compared to LO NRQCD calculations [38] (*right*)

## 2.2.2   Quarkonium Production in the Pre-LHC Era

The history of quarkonium physics has been accompanied by several long-standing problems, including several experimental inconsistencies. At this point, it is useful to illustrate in a few paragraphs the chronological developments in the field of quarkonium production, in order to understand its status in the pre-LHC era, and to appreciate the level of progress made thanks to the excellent performance of the LHC experiments and the simultaneous progress in the context of NRQCD.

The discovery of the J/$\psi$ meson, simultaneously in $p + Be$ collisions at BNL [30] and in $e^+e^-$ collisions at SLAC [31], was the first experimental sign of the charm quark, and opened up a new field in particle physics research, which is still actively followed in the LHC era.

Few years after the discovery, the intuitive and simple CSM was developed [32–34], with the ability to calculate quarkonium cross sections, differential in transverse momentum, without the need for free parameters that have to be fit to the data, as is the case in NRQCD. This CSM was first challenged by measurements of J/$\psi$ and $\psi(2S)$ cross sections by fixed-target experiments at Fermilab [35, 36], which exceeded the CSM predictions (at LO, at that time) by large factors. This was not regarded as a serious problem, especially for the J/$\psi$, which is affected by the – at that time – completely unknown feed-down fractions from heavier charmonium states. These measurements were made at relatively low values of $p_T$, where non-perturbative effects were also expected to justify differences between the data and the CSM curves.

The situation changed after the first measurements of quarkonium production cross sections at the Tevatron experiments, which could access much higher values of $p_T$ than previously explored, entering regions where $p_T \gg M_Q$, with $M_Q$ the mass of the quarkonium state $\mathcal{Q}$. The measured production cross sections of the prompt $\psi(nS)$, $\Upsilon(nS)$ [37, 39] and $\chi_c$ [40] quarkonia were once again large factors above the LO CSM calculations, as can be seen, in the case of the $\psi(nS)$, in the left panel of Fig. 2.6. At this point, the scientific community was alarmed, especially due to the $\psi(2S)$ discrepancy, which could not be attributed to any feed-down decays, the CSM underestimating the observed yields by a factor of 40–50, a problem also known as the "$\psi(2S)$ anomaly".

Roughly at that time, the NRQCD factorization approach was born [16]. The SDCs were first calculated at LO, and tested on Tevatron quarkonium cross section data. The $\psi(2S)$ anomaly could be successfully solved by adding CO contributions on top of the CS calculations, with free fit parameters representing the LDMEs of the individual CO contributions. The resulting fit could nicely describe the CDF $\psi(2S)$ data, as can be appreciated in the right panel of Fig. 2.6.

Despite the success of the LO NRQCD calculations, the trust in these new calculations was limited, given the enormous freedom of the model, with the overall normalization of the production cross section given by free parameters of the fit, and the shape of the distribution given by the relative importance of the individual color channels, characterized by $p_T$-distributions of different slopes. The obvious next step was to predict other measurable observables within the framework of NRQCD, and to measure them experimentally. The LDMEs fitted from the cross section measurements can be used to predict the polarization of the inclusive sample, built from color channels with different polarizations, with relative weights proportional to the LDMEs. With this approach, several groups conducting LO NRQCD calculations predicted almost fully transverse polarization in the HX frame, especially at high $p_\perp$, for Tevatron J/$\psi$ and $\psi(2S)$ production (see Ref. [41] and references therein). However, the CDF Collaboration measured no large polarizations [42, 43]. The CDF results for prompt J/$\psi$ and $\psi(2S)$ polarizations [43] are compared to the LO NRQCD calculations from Ref. [41] in Fig. 2.7. The J/$\psi$ prediction for $\lambda_\vartheta$ includes feed-down effects from the $\psi(2S)$ and $\chi_{cJ}$ states, and can therefore be directly compared to the prompt measurement of the CDF Collaboration. The measurements of both the J/$\psi$ and $\psi(2S)$ polarizations are in clear disagreement with the LO NRQCD calculations, challenging their validity.

## Puzzles and Solutions

The disagreement of the LO NRQCD calculations and the Tevatron quarkonium polarization measurements was often referred to as the "quarkonium polarization puzzle", which received a lot of attention by the scientific community, and several approaches were tested to solve the problem. One attempt was to extend the LO calculations of NRQCD to NLO, a task performed by several groups [19, 22–27].

The analysis described in Ref. [44] was among the first to use NLO NRQCD calculations to attempt a fit to extract the LDMEs for J/$\psi$ production. Production cross section measurements from both hadroproduction, including early LHC measurements, and photoproduction at HERA were studied to extract the LDMEs, which were then used to predict the J/$\psi$ polarization. More details about this analysis – and several other similar NRQCD analyses – can be found in Sect. 5.2. The analysis was

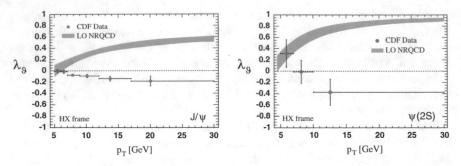

**Fig. 2.7** Prompt $J/\psi$ (*left*) and $\psi(2S)$ (*right*) polarization parameter $\lambda_{\vartheta}$ measured in the HX frame by the CDF Collaboration [43], compared to LO NRQCD calculations [41]

a success in the sense that the production cross sections of different collision systems could be reproduced simultaneously, hinting at the validity of the universality claim of the NRQCD factorization approach. However, the predicted $J/\psi$ polarization was very similar to the LO predictions, almost fully transverse in the HX frame, especially at high $p_T$. Therefore, the CDF polarization measurements could not be explained by these NLO NRQCD calculations.

The large differences between the CS LO and NLO calculations opened the possibility that the full CS quarkonium cross sections could be large enough to describe the data (or ensure that only a rather small CO component is needed). The NNLO* calculations described above were compared to the available data, for both cross section and polarization measurements [28, 45]. The $J/\psi$ and $\Upsilon(nS)$ cross sections for CS NNLO* calculations do not describe the data, which are systematically above the calculations. The CS NNLO* polarizations are similar to the CS NLO calculations, but slightly more longitudinal. Therefore, these calculations cannot describe the mostly unpolarized CDF $J/\psi$ data. However, this comparison is not entirely fair given that the CS NNLO* calculations do not include feed-down decays. Nevertheless, the CS NNLO* calculations supported the idea that CO contributions are indeed necessary to explain quarkonium production data.

One further unfortunate but important component of the quarkonium polarization story of the pre-LHC era are a series of experimental inconsistencies. The Tevatron experiments have published results for quarkonium polarization which are mutually inconsistent, as illustrated in Fig. 2.8. The left panel shows CDF measurements of the prompt $J/\psi$ polarization in the HX frame from data taken in different run periods [42, 43]. The slight change in CM energy from $\sqrt{s} = 1.8$ TeV to $\sqrt{s} = 1.96$ TeV, as well as the slight difference in the rapidity regions of the measurements cannot explain the large differences among the results. The right panel shows measurements of the $\Upsilon(1S)$ polarization in the HX frame as measured at $\sqrt{s} = 1.96$ TeV by the CDF Collaboration [46] and the D0 Collaboration [47]. Also in this case one can see a large discrepancy, which again cannot be explained by the different rapidity regions of the measurements.

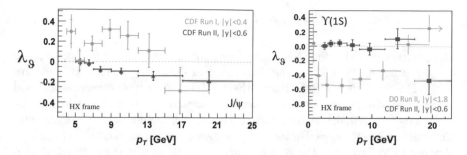

**Fig. 2.8** Measurements of the polarization parameter $\lambda_\vartheta$ in the HX frame, for the prompt $J/\psi$ by the CDF Collaboration [42, 43] (*left*) and for the $\Upsilon(1S)$ by the CDF Collaboration [46] and the D0 Collaboration [47] (*right*)

Given that the data could not be fully trusted, the disagreement with the LO NRQCD calculations was not regarded as a major problem in the pre-LHC era. These unfortunate inconsistencies are the main reason why quarkonium polarization measurements were not considered in the NRQCD fits of the LDMEs, but rather used as a check to compare the NRQCD predictions for the polarization observables, which are obtained a-posteriori. This is especially inauspicious as the polarization observables, given the clear differences between the individual color channels, give more intuitive information about the relative importance of the channels than the differential cross sections, which are relatively similar for all underlying processes.

At this point, it can only be speculated why the Tevatron era was affected by such inconsistencies. Due to progress in the understanding of quarkonium polarization (see Sect. 2.3), it is now clear that 1-dimensional angular measurements, only considering the polar anisotropy $\lambda_\vartheta$ in one frame, as was the strategy of these Tevatron measurements, leads to major problems. Besides the ambiguity of results only stating $\lambda_\vartheta$ (see Sect. 2.3.3 for more details), there are experimental pitfalls that can introduce large biases when integrating over the azimuthal component of the decay [48].

## 2.3 Quarkonium Polarization

Driven by the observed inconsistencies of the quarkonium polarization measurements at the Tevatron experiments, there has been huge progress in the understanding of quarkonium polarization and in the methodology required for the corresponding measurements. This progress was documented in a series of papers [48–53]. Some general considerations and the most important findings are discussed here. All quarkonium polarization measurements conducted at hadron colliders since the development of this new understanding follow this methodology, with the pleasant consequence that these new results show a consistent picture throughout various experiments (see Sect. 5.1.2).

## 2.3.1  General Considerations

This part is restricted to the discussion of the polarization of vector $1^{--}$ states in the dilepton decay, $Q^{3}S_{1} \to l^{+}l^{-}$. The polarization of the $J^{++}$ $\chi$ states in the radiative decays $Q^{3}P_{J} \to Q^{3}S_{1} + \gamma$ is discussed in Sect. 2.3.4. A vector particle can be observed in three eigenstates of the angular momentum component $J_{z}$, with respect to a quantization axis $z$, $J_{z} = 0, \pm 1$. If, on a statistical basis, quarkonium states are dominantly observed in either the $J_{z} = 0$ or the $J_{z} = \pm 1$ eigenstates, the state is called polarized, with respect to the axis $z$. The polarization of a quarkonium state can be interpreted as the preferred spin alignment, which can be caused by basic conservation laws and symmetries of the electroweak and strong interactions, depending on the properties of the corresponding production diagrams. A preferred spin alignment affects the decay angular distribution of the two leptons, in the quarkonium rest-frame, and can therefore be measured from this distribution. An isotropic angular decay distribution corresponds to unpolarized quarkonia, while anisotropies of the distribution reflect a polarized state. In case of preferred spin alignment corresponding to the projections $J_{z} = \pm 1$, the quarkonium is denoted as transversely polarized, in case of a preferred projection of $J_{z} = 0$, the quarkonium is denoted as longitudinally polarized. In case the quarkonium state is produced exclusively in either $J_{z} = \pm 1$ or $J_{z} = 0$, the polarization is denoted as fully transverse or fully longitudinal, respectively.

The angular distribution is measured with respect to a polarization reference frame, in the quarkonium rest frame. The definitions of the polar angle $\vartheta$ and azimuthal angle $\varphi$ are shown in Fig. 2.9 (top). The decay angles are defined as the angles of the positive lepton with respect to the reference frame, whose $x$-$z$ plane is defined by the production plane (bottom left), built by joining the momentum vector of the quarkonium state with the momentum vector of the colliding beams ($\vec{b}_{1}^{c}$ and $\vec{b}_{2}^{c}$), in the laboratory frame. The $y$ axis is defined to be perpendicular to the production plane, in the direction of $\vec{b}_{1}^{r} \times \vec{b}_{2}^{r}$ and $\vec{b}_{2}^{r} \times \vec{b}_{1}^{r}$, with $\vec{b}_{1}^{r}$ and $\vec{b}_{2}^{r}$ defined in the quarkonium rest frame, for positive and negative rapidities, respectively.

The reference frame is then fully defined by choosing a quantization axis $z$ within the production plane. This choice can in principle be done arbitrarily, but there are some physically motivated choices for the quantization axis with respect to which the polarization is measured, whose definitions are shown in the bottom right panel of Fig. 2.9. The HX axis is defined to be aligned with the quarkonium flight direction. The Collins-Soper axis [54] is defined as the opposite direction of the bisector of the two momentum vectors of the colliding beams, which is an approximation of the direction of the colliding partons. The third definition considered in this thesis is the Perpendicular-Helicity (PX) axis, defined to be orthogonal to the Collins-Soper axis. The definitions of the individual frames depend on the quarkonium production kinematics. In the limit of high $p_{T}$ and mid-rapidity, the HX and Collins-Soper frames are orthogonal, in which case the PX and HX frames are identical. In the opposite limit of $p_{T} \to 0$ and forward-rapidity, the HX and Collins-Soper frames are identical, and the HX and PX frames are orthogonal. The usage of the PX frame ensures, independently of the kinematical region of a measurement, that two orthogonal frames

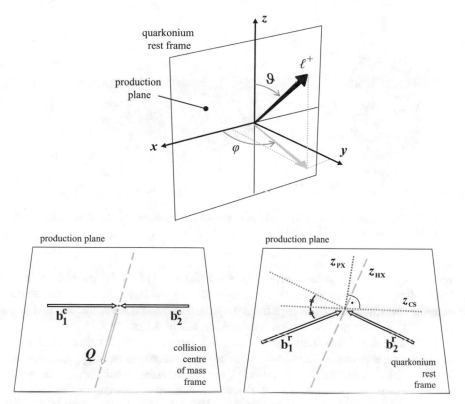

**Fig. 2.9** Definitions of the polar angle $\vartheta$ and of the azimuthal angle $\varphi$ of the polarization reference frame (*top*), of the production plane (*bottom left*) and of the quantization axis $z$ (*bottom right*) [48]

can always be considered in the analysis, which is an important requirement for a reliable measurement, as argued in Ref. [49].

The angular decay distribution of a vector state can be calculated from basic quantum mechanical considerations, requiring helicity conservation at the photon-dilepton vertex of the $Q^{3S_1} \to l^+l^-$ decay. The most general angular decay distribution of a parity-conserving dilepton decay of a vector particle can be written as [48]

$$W(\cos\vartheta, \varphi|\vec{\lambda}) \propto \frac{1}{(3+\lambda_\vartheta)}(1 + \lambda_\vartheta \cos^2\vartheta +$$
$$+ \lambda_\varphi \sin^2\vartheta \cos 2\varphi + \lambda_{\vartheta\varphi} \sin 2\vartheta \cos\varphi) . \tag{2.5}$$

This distribution is parametrized by three "anisotropy parameters" $\vec{\lambda} = (\lambda_\vartheta, \lambda_\varphi, \lambda_{\vartheta\varphi})$, also referred to as the "polarization parameters". The parameter $\lambda_\vartheta$ describes the polar anisotropy of the decay, $\lambda_\varphi$ describes the azimuthal anisotropy of the decay, and $\lambda_{\vartheta\varphi}$ describes the change of the azimuthal anisotropy as a function of the polar angle $\vartheta$. The polar anisotropy parameter $\lambda_\vartheta$ is positive (negative) in case of trans-

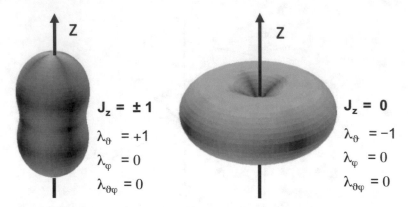

$J_z = \pm 1$

$\lambda_\vartheta = +1$

$\lambda_\varphi = 0$

$\lambda_{\vartheta\varphi} = 0$

$J_z = 0$

$\lambda_\vartheta = -1$

$\lambda_\varphi = 0$

$\lambda_{\vartheta\varphi} = 0$

**Fig. 2.10** Fully transverse (*left*) and fully longitudinal (*right*) decay angular distributions, with respect to the quantization axis $z$ [48]

verse (longitudinal) polarization, and $\lambda_\vartheta = +1$ ($\lambda_\vartheta = -1$) for fully transverse (fully longitudinal) vector states. Figure 2.10 shows the angular distributions for these two extreme cases, where the distance from the origin to the surface corresponds to the probability that the positive lepton is emitted in this direction.

If the angular distribution results from $n$ samples $i$ of vector quarkonia with different angular distributions parametrized by different anisotropy parameters, $W(\cos\vartheta, \varphi | \vec{\lambda}^{(i)})$, with relative weights $f^{(i)}$, the total angular distribution can be written as the sum $\sum_{i=1}^n f^{(i)} W(\cos\vartheta, \varphi | \vec{\lambda}^{(i)})$, and the effective polarization parameters, describing the inclusive angular distribution, $\vec{\lambda}'$, can be calculated as [48]

$$\vec{\lambda}' = \frac{\sum_{i=1}^n \frac{f^{(i)}}{3+\lambda_\vartheta^{(i)}} \vec{\lambda}^{(i)}}{\sum_{i=1}^n \frac{f^{(i)}}{3+\lambda_\vartheta^{(i)}}} . \tag{2.6}$$

This "polarization sum rule" is important for combining different angular distributions of feed-down decays, as well as for the addition of the polarizations of the individual color channels in NRQCD calculations. It is important to note that, in the case of several contributions characterized by different polarizations, longitudinal components carry a "heavier weight" than the transverse components. As a simple example, considering a mixture of a fully longitudinal component and a fully transverse component, with equal weights $f^{(i)} = 0.5$, the resulting effective polar anisotropy of the sum of the components is $\lambda'_\vartheta = -1/3$, very different from the unpolarized distribution that intuition might suggest.

### 2.3.2 Frame-Invariant Formalism

It is useful to introduce the concept of the "natural polarization frame". For any decay angular distribution of a vector quarkonium state one can define a reference frame in which the polar anisotropy is maximal ($\lambda_\vartheta^{nat}$), and in which $\lambda_\varphi$ is minimal and $\lambda_{\vartheta\varphi}$ vanishes [50]. This frame is denoted as the natural polarization frame for the given angular distribution. Each process contributing to quarkonium production, for which $\lambda_\vartheta^{nat} \neq 0$, has a natural polarization axis. With the condition that $\lambda_\vartheta \in [-1, 1]$ with respect to any $z$ axis (equivalent to the condition that the natural polar anisotropy $\lambda_\vartheta^{nat} \in [-1, 1]$) one can derive so-called "positivity constraints" for $\lambda_\varphi$ and $\lambda_{\vartheta\varphi}$ [52], which constrain the allowed phase space of $\vec{\lambda}$ to the regions shown as the grey areas in Fig. 2.11.

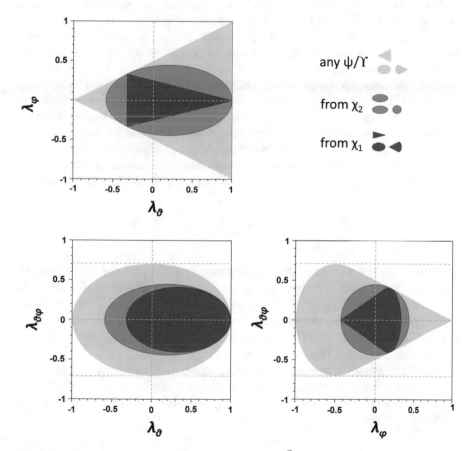

**Fig. 2.11** Allowed phase space regions of the parameters $\vec{\lambda}$ for $1^{--}$ (*grey*) and $J^{++}$ quarkonia, with J = 1, 2 (*dark* and *light blue*, respectively) [53]

The relative importance of the fundamental processes of quarkonium production can change as a function of the quarkonium kinematics, given the different $p_T$-slopes of the SDCs as discussed in Sect. 2.2.1, and the feed-down decays that might have varying relative importance as a function of the kinematics. Due to the different angular distributions of the individual processes, one expects a kinematic dependence of the inclusive set of parameters $\vec{\lambda}$, which is denoted here as "intrinsic" kinematic dependence. However, as the parameters $\vec{\lambda}$ are frame-dependent, and as the definition of the frames depends on the kinematics, the measured polarization in a given reference frame can show a kinematic dependence simply because the measurement frame is not the natural frame, especially in cases of mixtures of processes characterized by different natural polarization axes. This dependence is denoted here as "extrinsic", and is an artifact of the measurement, not reflecting a real physics effect, but purely kinematical effects.

Even though the parameters $\vec{\lambda}$ are frame-dependent, the shape of the angular distribution is independent of the chosen reference frame. It is possible to define frame-invariant observables [50], as combinations of the frame-dependent parameters $\vec{\lambda}$. These frame-invariant quantities do not depend on the frame in which the polarization is measured, and therefore allow us to differentiate between intrinsic and extrinsic kinematic dependencies – between physics effects and purely kinematical effects. These quantities therefore carry viable physical information about the nature of the polarization of a given quarkonium state. The combination of frame-dependent and frame-invariant observables is vital, given that the frame-invariant quantities do not carry any information about the natural polarization frames.

There are an infinite number of frame-invariant observables [50] that can be defined. The most convenient parameter, widespread in the literature, is $\tilde{\lambda}$, defined as

$$\tilde{\lambda} = \frac{\lambda_\vartheta + 3\lambda_\varphi}{1 - \lambda_\varphi} \ . \tag{2.7}$$

This parameter is $+1$ for any fully transverse shape and $-1$ for any fully longitudinal shape. This means, for example, that for any mixture of processes where each component is characterized by fully transverse polarization with respect to different natural polarization axes, the resulting measurement of the inclusive sample in any frame will nevertheless result in $\tilde{\lambda} = +1$. Given the allowed ranges for $\lambda_\vartheta$ and $\lambda_\varphi$, $\tilde{\lambda}$ is contained within the interval $[-1, \infty]$.

Besides its physics information, the measurement of $\tilde{\lambda}$ provides a critical experimental cross check. The comparison of measurements of $\tilde{\lambda}$ in several frames, including at least two orthogonal frames, can reveal systematic biases that are not accounted for in the analysis.

### 2.3.3 Ambiguity of Pre-LHC Quarkonium Polarization Measurements

The importance of following this methodology can be shown on the basis of the prompt $J/\psi$ polarization analysis from CDF [43], which is shown in the left panel of Fig. 2.7. This measurement has been conducted before the development of the methodology as described in this section. Consequentially, the analysis was performed integrating over the azimuthal component of the decay, only providing the parameter $\lambda_\vartheta$, only in the HX frame, neither providing information about the azimuthal anisotropy, nor providing frame-invariant information through $\tilde{\lambda}$, which was not known at the time of the measurement.

Given the unmeasured azimuthal anisotropy $\lambda_\varphi$, this measurement allows for several very different physical interpretations, leading to an ambiguity that can only be resolved by further measurements of the prompt $J/\psi$ polarization, following the recipes summarized in this section. In order to visualize the ambiguity of this measurement, Ref. [55] introduces three polarization scenarios which are all compatible with the CDF measurement of $\lambda_\vartheta$, but whose 2-dimensional angular decay distributions are considerably different. The first (second) scenario assumes that the $J/\psi$ decay angular distribution does not reveal any azimuthal anisotropy in the HX (Collins-Soper) frame, $\lambda_\vartheta^{HX} = 0$ ($\lambda_\vartheta^{CS} = 0$). The third scenario assumes a certain fraction (slightly changing as a function of $p_T$) of all $J/\psi$'s to be produced transversely polarized with respect to the HX frame, the rest transversely polarized with respect to the Collins-Soper frame.

Figure 2.12 shows the kinematic dependence of the polarization parameters in the HX frame, for these three scenarios. This figure displays the measured parameter $\lambda_\vartheta$, showing that the individual scenarios are almost identical in $\lambda_\vartheta$, as well as the unmeasured parameters $\lambda_\varphi$, $\lambda_{\vartheta\varphi}$ and $\tilde{\lambda}$.

This pedagogical example illustrates that restricting a measurement to the polar anisotropy $\lambda_\vartheta$ does not provide the necessary information to interpret the polarization of a quarkonium state, as very different physical scenarios can be compatible with such a measurement. However, if one measures also the azimuthal component and the frame-invariant parameter $\tilde{\lambda}$, the individual scenarios can be very easily distinguished. These calculations are further provided for the accessible rapidity ranges of the CMS detector as well as of the LHCb detector [55], allowing for an a-posteriori interpretation of the CDF results.

### 2.3.4 Polarization of the $\chi$ States

In principle, the polarizations of the P-wave quarkonia are more complicated to measure than the polarizations of S-wave quarkonia. The $\mathcal{Q}^{3P_J} \rightarrow \mathcal{Q}^{3S_1} + \gamma$ decays are much more challenging to reconstruct due to the presence of a low energy photon, difficult to detect and reconstruct. The direct way to measure P-wave polarization would be to measure the angular distribution of the $\mathcal{Q}^{3S_1} + \gamma$ system with respect to a reference frame in the $\mathcal{Q}^{3P_J}$ rest frame. This is very challenging, as the photon

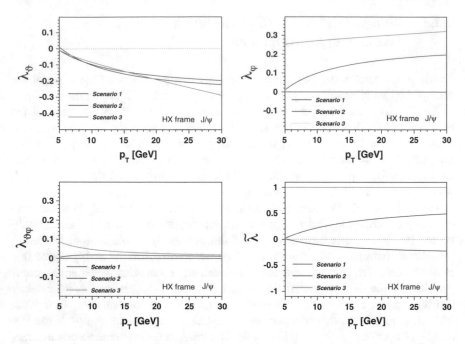

**Fig. 2.12** Kinematic behavior of anisotropy parameters $\lambda_\vartheta$, $\lambda_\varphi$, $\lambda_{\vartheta\varphi}$ in the HX frame and $\tilde{\lambda}$, in the rapidity region $|y| < 0.6$, for the three scenarios as discussed in the text [55]

kinematics and efficiencies have to be accurately known for such studies. However, it was shown recently [53] that the $Q^{3P_J}$ angular distribution in the $Q^{3P_J}$ rest frame can be very well approximated by the dilepton angular distribution of the $Q^{3S_1} \rightarrow l^+l^-$ decay in the $Q^{3S_1}$ rest frame, for sufficiently large momenta of the original $Q^{3P_J}$ system. For the momenta measurable by the CMS detector, the bias associated to this approximation is negligible. Therefore, the measurement of the $J^{++}$ states can be done in the exact same way as the measurement of the $1^{--}$ states, except for identifying those $1^{--}$ states that are accompanied by a photon originating from the dilepton vertex, with an invariant mass close to the mass of the $J^{++}$ state under study.

Similarly to the allowed regions for $1^{--}$ states, one can derive the allowed regions for $\tilde{\lambda}$ also for the $J^{++}$ states, shown in Fig. 2.11 as the dark and light blue areas, respectively for J=1 and J=2. The interpretation of the polarization of $J^{++}$ states in terms of quantum states and preferred spin alignment is not as intuitive as the polarization of the $1^{--}$ states. The $1^{++}$ states have possible eigenstates corresponding to projections of the angular momentum on the z-axis of $J_z = 0, \pm1$, while the $2^{++}$ states can have the projections $J_z = 0, \pm1, \pm2$. The fully transverse polar anisotropy $\lambda_\vartheta = +1$ corresponds to the $J_z = 0$ and $J_z = \pm2$ eigenstates for J=1 and J=2 states, respectively. The partially longitudinal polar anisotropy $\lambda_\vartheta = -1/3$ corresponds to the $J_z = \pm1$ eigenstates, for both J=1 and J=2 states, while the $J_z = 0$ eigenstate of the J=2 states corresponds to the minimum polar anisotropy of $\lambda_\vartheta = -3/5$ [53].

## 2.4 Quarkonium Physics Summary

In this chapter the spectra of the charmonium and bottomonium meson families were introduced, followed by a discussion about the theoretical state-of-the-art framework describing quarkonium production, the NRQCD factorization approach, and a review of the situation in the field of quarkonium production physics in the pre-LHC era. The Tevatron results clarified that the color-singlet contributions cannot be solely responsible for quarkonium production in hadron collisions, and that color-octet transitions are realized in nature. However, the confusion in the quarkonium physics community was large at that time, mostly due to experimental inconsistencies in quarkonium polarization measurements. It is clear that the community eagerly awaited "better data", and therefore an experimental clarification from the LHC experiments, especially providing quarkonium polarization data with improved and more robust analysis techniques. The LHC quarkonium physics program does not only aim at clarifying the experimental situation of quarkonium polarization, but also at extending the $p_T$ reach of the measurements, for both production cross section measurements as well as polarization measurements, for S-wave and P-wave states, far beyond the reach of the Tevatron experiments.

The measurements at the core of this thesis have been motivated by introducing the theoretical foundation of quarkonium production physics and the existing experimental problems. The interplay of progress in theory and experiment through the LHC programs and beyond is necessary to understand the processes that lead to quarkonium production, addressing the basic and general question of how quarks bind into hadrons via the strong force.

## References

1. Olive KA et al (2014) Particle data group. Chin Phys C 38:090001
2. LHCb Collaboration (2014a) Study of $\chi_b$ meson production in $pp$ collisions at $\sqrt{s} = 7$ and 8 TeV and observation of the decay $\chi_b(3P) \to \Upsilon(3S)\gamma$. Eur Phys J C 74:3092
3. LHCb Collaboration (2014b) Measurement of the $\chi_b(3P)$ mass and of the relative rate of $\chi_{b1}(1P)$ and $\chi_{b2}(1P)$ production. J High Energy Phys 1410:88
4. ATLAS Collaboration (2012) Observation of a new $\chi_b$ state in radiative transitions to $\Upsilon(1S)$ and $\Upsilon(2S)$ at ATLAS. Phys Rev Lett 108:1520012012
5. D0 Collaboration (2012) Observation of a narrow mass state decaying into $\Upsilon(1S) + \gamma$ in $p\bar{p}$ collisions at $\sqrt{s} = 1.96$ TeV. Phys Rev D 86:031103
6. QWG Collaboration (2011) Heavy quarkonium: progress, puzzles, and opportunities. Eur Phys J C 71:1534
7. Belle Collaboration (2003) Observation of a narrow charmonium - like state in exclusive $B^\pm \to K^\pm \pi^+ \pi^- J/\psi$ decays. Phys Rev Lett 91:262001
8. LHCb Collaboration (2013) Determination of the X(3872) meson quantum numbers. Phys Rev Lett 110:222001
9. Belle Collaboration (2008) Observation of a resonance-like structure in the $\pi^\pm \psi'$ mass distribution in exclusive $B \to K\pi^+ \psi'$ decays. Phys Rev Lett 100:142001
10. LHCb Collaboration (2014) Observation of the resonant character of the $Z(4430)^-$ state. Phys Rev Lett 112:222002

11. Gell-Mann M (1964) A schematic model of baryons and mesons. Phys Lett 8:214
12. Matsui T, Satz H (1986) $J/\psi$ suppression by quark-gluon plasma formation. Phys Lett B 178:416
13. Digal S, Petreczky P, Satz H (2001) Quarkonium feed-down and sequential suppression. Phys Rev D 64:094015
14. Braun-Munzinger P, Stachel J (2000) (Non)thermal aspects of charmonium production and a new look at $J/\psi$ suppression. Phys Lett B 490:196
15. Bodwin GT, Petriello F, Stoynev S, Velasco M (2013) Higgs boson decays to quarkonia and the $H\bar{c}c$ coupling. Phys Rev D 88:053003
16. Bodwin GT, Braaten E, Lepage GP (1995) Rigorous QCD analysis of inclusive annihilation and production of heavy quarkonium. Phys Rev D 51:1125
17. Fritzsch H (1977) Producing heavy quark flavors in hadronic collisions – 'A test of quantum chromodynamics'. Phys Lett B 67:217
18. Private communication from P. Faccioli
19. Butenschön M, Kniehl B (2012) $J/\psi$ polarization at Tevatron and LHC: Nonrelativistic-QCD factorization at the crossroads. Phys Rev Lett 108:172002
20. Schuler GA (1997) Quarkonium production: Velocity scaling rules and long distance matrix elements. Int J Mod Phys A 12:3951
21. Price DD (2008) Studies of quarkonium production and polarisation with early data at ATLAS. Ph.D. thesis, Lancaster University
22. Artoisenet P, Lansberg JP, Maltoni F (2007) Hadroproduction of $J/\psi$ and $\Upsilon$ in association with a heavy-quark pair. Phys Lett B 653:60
23. Campbell JM, Maltoni F, Tramontano F (2007) QCD corrections to $J/\psi$ and Upsilon production at hadron colliders. Phys Rev Lett 98:252002
24. Gong B, Wang JX (2008) Next-to-leading-order QCD corrections to $J/\psi$ polarization at Tevatron and Large-Hadron-Collider energies. Phys Rev Lett 100:232001
25. Gong B, Wang JX (2008) QCD corrections to polarization of $J/\psi$ and $\Upsilon$ at Tevatron and LHC. Phys Rev D 78:074011
26. Gong B, Li XQ, Wang JX (2009) QCD corrections to $J/\psi$ production via color octet states at Tevatron and LHC. Phys Lett B 673:197
27. Ma YQ, Wang K, Chao KT (2011) QCD radiative corrections to $\chi_{cJ}$ production at hadron colliders. Phys Rev D 83:111503
28. Artoisenet P, Campbell JM, Lansberg JP et al (2008) $\Upsilon$ production at Fermilab Tevatron and LHC energies. Phys Rev Lett 101:152001
29. Lansberg JP (2009) Real next-to-next-to-leading-order QCD corrections to $J/\psi$ and Upsilon hadroproduction in association with a photon. Phys Lett B 679:340
30. Aubert JJ et al (1974) Experimental observation of a heavy particle $J$. Phys Rev Lett 33:1404
31. Augustin JE et al (1974) Discovery of a narrow resonance in $e^+e^-$ annihilation. Phys Rev Lett 33:1406
32. Kartvelishvili VG, Likhoded AK, Slabospitsky SR (1978) $D$ meson and $\psi$ meson production in hadronic interactions. Sov J Nucl Phys 28:678
33. Baier R, Ruckl R (1983) Hadronic collisions: a quarkonium factory. Z Phys C 19:251
34. Glover EW, Martin AD, Stirling WJ (1988) $J/\psi$ production at large transverse momentum at hadron colliders. Z Phys C 38:473
35. E789 Collaboration (1995) Measurement of $J/\psi$ and $\psi'$ production in 800-GeV/c proton-gold collisions. Phys Rev D 52:1307
36. McGaughey PL (1996) Recent measurements of quarkonia and Drell-Yan production in proton nucleus collisions. Nucl Phys A 610:394C
37. CDF Collaboration (1997) $J/\psi$ and $\psi(2S)$ production in $p\bar{p}$ collisions at $\sqrt{s} = 1.8$ TeV. Phys Rev Lett 79:572
38. Krämer M (2001) Quarkonium production at high-energy colliders. Progr Part Nucl Phys 47:141
39. CDF Collaboration (1996) Quarkonia production at CDF. Nucl Phys A 610:373C

40. CDF Collaboration (1997) Production of $J/\psi$ mesons from $\chi_c$ meson decays in $p\bar{p}$ collisions at $\sqrt{s} = 1.8$ TeV. Phys Rev Lett 79:578
41. Braaten E, Kniehl B, Lee J (2000) Polarization of prompt $J/\psi$ at the Tevatron. Phys Rev D 62:094005
42. CDF Collaboration (2000) Measurement of $J/\psi$ and $\psi(2S)$ polarization in $p\bar{p}$ collisions at $\sqrt{s} = 1.8$ TeV. Phys Rev Lett 85:2886
43. CDF Collaboration (2007) Polarization of $J/\psi$ and $\psi(2S)$ mesons produced in $p\bar{p}$ collisions at $\sqrt{s} = 1.96$ TeV. Phys Rev Lett 99:132001
44. Butenschön M, Kniehl B (2012) $J/\psi$ production in NRQCD: a global analysis of yield and polarization. Nucl Phys B Proc Suppl 151:222–224
45. Lansberg JP (2011) $J/\psi$ production at $\sqrt{s} = 1.96$ and 7 TeV: color-singlet model, NNLO* and polarisation. arXiv:1107.0292
46. CDF Collaboration. CDF public note 9966
47. D0 Collaboration (2008) Measurement of the polarization of the $\Upsilon(1S)$ and $\Upsilon(2S)$ states in $p\bar{p}$ collisions at $\sqrt{s} = 1.96$ TeV. Phys Rev Lett 101:182004
48. Faccioli P, Lourenço C, Seixas J, Wöhri H (2010) Towards the experimental clarification of quarkonium polarization. Eur Phys J C 69:657
49. Faccioli P, Lourenço C, Seixas J, Wöhri H (2009) $J/\psi$ polarization from fixed-target to collider energies. Phys Rev Lett 102:151802
50. Faccioli P, Lourenço C, Seixas J (2010a) Rotation-invariant relations in vector meson decays into fermion pairs. Phys Rev Lett 105:061601
51. Faccioli P, Lourenço C, Seixas J (2010b) A new approach to quarkonium polarization studies. Phys Rev D 81:111502
52. Faccioli P, Lourenço C, Seixas J, Wöhri H (2011a) Model-independent constraints on the shape parameters of dilepton angular distributions. Phys Rev D 83:056008
53. Faccioli P, Lourenço C, Seixas J, Wöhri H (2011b) Determination of $\chi_c$ and $\chi_b$ polarizations from dilepton angular distributions in radiative decays. Phys Rev D 83:096001
54. Collins JC, Soper DE (1977) Angular distribution of dileptons in high-energy hadron collisions. Phys Rev D 16:2219
55. Knünz V (2011) Measurement of $J/\psi$ polarization with the CMS experiment in proton-proton collisions at $\sqrt{s} = 7$ TeV. Diploma Thesis, Institute of High Energy Physics, Vienna University of Technology

# Chapter 3
# Experimental Setup

This chapter is dedicated to introducing the experimental setup used to obtain the results discussed in Chap. 4, which are based on data collected in $pp$ collisions at the CMS detector in the years 2011 and 2012. The LHC is briefly introduced in Sect. 3.1, followed by a more detailed introduction of the CMS experiment in Sect. 3.2, highlighting the parts of the detector, data acquisition chain and reconstruction software that are relevant for the analyses discussed in this thesis, based on the $\mu\mu$ final state. No generic details shall be discussed here, all relevant additional information can be found in the extensive documentation describing the LHC [1] and the CMS detector [2–4].

## 3.1 The Large Hadron Collider

### 3.1.1 The Machine

The LHC is a superconducting two-ring synchrotron accelerator and collider, which performs at collision rates and collision energies that are far beyond previous hadron colliders. It is located in a tunnel with 27 km circumference, 45–170 m below ground, at CERN. The machine is equipped with the possibility to accelerate both protons and heavy ions, and has so far provided $pp$, proton-lead and lead-lead collisions. The results of this thesis are based on data collected with $pp$ collisions, to which the following remarks are limited to.

Protons are not accelerated in a continuous way, but in "packages", so-called "proton bunches". These bunches are pre-accelerated using the CERN accelerator complex as shown in Fig. 3.1, including a proton source, a linear accelerator, and several circular accelerators. The proton bunches are finally injected into the LHC rings at an energy of 450 GeV per proton. The total time needed for the

© Springer International Publishing Switzerland 2017
V. Knünz, *Measurement of Quarkonium Polarization to Probe QCD at the LHC*, Springer Theses, DOI 10.1007/978-3-319-49935-2_3

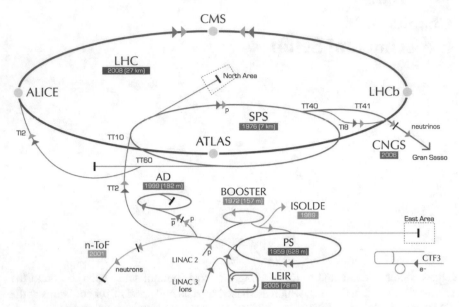

**Fig. 3.1** The CERN accelerator complex [5]

pre-acceleration until the proton beams are injected into the LHC is approximately four minutes [1]. The protons are further accelerated in the LHC with a supercon-ducting cavity system operating at 400 MHz. The proton bunches are kept on the approximately circular path with superconducting dipole magnets operating at a field of up to 8.33 T, while quadrupole magnets are used to focus the proton beams. Once the two proton beams are accelerated from 450 GeV to the chosen energy, which takes approximately 20 min for the nominal design value of 7 TeV per beam, they are brought to collision in the center of the LHC experiments. The beams are kept in this collision mode for several hours, until the beam intensity decreases below a certain threshold, due to the loss of protons and focus of the beams. At this point, the full cycle restarts with the injection of a new "LHC fill" [1].

In the 2010 and 2011 $pp$ runs each proton was accelerated up to 3.5 TeV (CM energy of $\sqrt{s} = 7$ TeV), in the 2012 run up to 4 TeV ($\sqrt{s} = 8$ TeV), and after the long shutdown period in 2013/2014, the LHC will restart $pp$ collisions in 2015 with 6.5 TeV per proton ($\sqrt{s} = 13$ TeV).[1] The nominal design $\sqrt{s}$ for $pp$ collisions at the LHC is 14 TeV, limited by the strength of the dipole magnets.

The LHC has four main collision points, located at the main experiments, the two general purpose experiments CMS and ATLAS (A Toroidal LHC ApparatuS), as well as LHCb (LHC beauty), mostly devoted to b-physics measurements, and ALICE (A Large Ion Collider Experiment), specialized in, but not limited to, the

---

[1] The run periods at 7 and 8 TeV are denoted as "Run I", while the run at 13 TeV starting in 2015 is referred to as "Run II".

analysis of data collected in HI collisions. The LHC layout and the location of the four experiments can be seen in Fig. 3.1.

The LHC nominal design foresees $1.15 \cdot 10^{11}$ protons per bunch, a maximum number of bunches of 2808, corresponding to individual bunches of 2.5 ns length, every 25 ns. These nominal design values are foreseen to be reached in 2015. In the running periods relevant for the analyses discussed here, the "bunch spacing" was 50 ns. At the end of 2012, the number of protons per bunch was around $1.55 \cdot 10^{11}$, exceeding the nominal design value [6]. To estimate the number of $pp$ collisions, the "instantaneous luminosity" is defined,

$$L = \frac{N_b^2 \, n_b \, f_{rev} \, \gamma_r}{4\pi \, \epsilon_n \, \beta^\star} \, F \,, \tag{3.1}$$

with $N_b$ the number of protons per bunch, $n_b$ the number of bunches per proton beam, $f_{rev}$ the revolution frequency, $\gamma_r$ the relativistic gamma factor, $\epsilon_n$ the normalized transverse beam emittance, $\beta^\star$ the betatron function at the collision point, and F the geometric luminosity reduction factor due to the beam crossing angle [1]. The LHC design instantaneous luminosity is $10^{34}$ cm$^{-2}$s$^{-1}$.[2] The actual luminosity delivered by the LHC in the years 2010–2012 is discussed in more detail in Sect. 3.2.1.

The CMS and ATLAS experiments are designed to be able to cope with the LHC nominal design luminosity, while LHCb and ALICE are limited to lower values. In 2012, the LHC delivered a factor of around 10 times less luminosity to LHCb, with respect to CMS, and a factor of more than 2 000 less luminosity to ALICE [6]. The ALICE event rate is limited by the readout time of their core tracking detector, the TPC, a slow detector, relatively speaking. In LHCb the limitation is justified by the requirement of a negligible probability of multiple pp collisions per "bunch crossing", a phenomenon denoted as "pile-up" (PU).

The number of particles that are produced in a certain process per second can be calculated by

$$N_p = L \cdot \sigma_p \,, \tag{3.2}$$

where $\sigma_p$ represents the production cross section for the process under consideration [1]. To estimate the total amount of collisions delivered by the LHC in a certain time period $\Delta t$, the instantaneous luminosity, $L(t)$, is integrated over time $t$,

$$\hat{L}_{\Delta t} = \int_{\Delta t} L(t) \mathrm{d}t \,, \tag{3.3}$$

with $\hat{L}_{\Delta t}$ denoted as the "integrated luminosity". The usual unit used for the integrated luminosity is "events per barn", written as b$^{-1}$. More reasonably, for the LHC $pp$ operations, the units pb$^{-1}$ (= $10^{12}$ b$^{-1}$) and fb$^{-1}$ (= $10^{15}$ b$^{-1}$) are used.

---

[2] $L = 10^{34}$ cm$^{-2}$s$^{-1}$ can also be written as 10 Hz/nb, with 1b= $10^{-24}$ cm$^2$.

### 3.1.2   Physics at the LHC

The physics objectives of the LHC program are very diverse. The main purpose was
defined to be the clarification of the nature of the electroweak symmetry breaking, for
which the Higgs mechanism was presumed to be responsible. Indeed, this endeavor
has proven to be successful, with the announcement of the discovery of a Higgs-like
boson on July $4^{th}$, 2012, by the CMS and ATLAS Collaborations [7, 8] with a mass
of around 125 GeV. All subsequent studies of the properties of this boson show that
it is compatible with the standard model Higgs boson.

Figure 3.2 gives an overview of SM cross sections as a function of $\sqrt{s}$ in $pp$ and
$p\bar{p}$ collisions, for several benchmark processes relevant for LHC physics studies,
including estimates of the expected number of produced particles for each process per
second, for a given luminosity of $10^{33}$ cm$^{-2}$s$^{-1}$. This information emphasizes clearly
the strategy of the LHC design to optimize the physics output of its experiments. The
nominal $\sqrt{s}$ was maximized in order to increase the cross sections of the processes
of interest, while at the same time the nominal luminosity was maximized in order
to increase the number of particles of interest produced in the $pp$ collisions.

As a quantitative example, considering the total inelastic $pp$ cross section at
$\sqrt{s} = 8$ TeV, $\sigma_{pp} = 69.4$ mb, and the maximum instantaneous luminosity in the
2012 run of around $L = 7.7 \cdot 10^{33}$ cm$^{-2}$s$^{-1}$ [9], the maximum average number of
collisions per second in 2012 was around $5.3 \cdot 10^8$, corresponding to an average of 26
collisions per bunch crossing. The LHC has delivered 23.3 fb$^{-1}$ integrated luminosity to the CMS interaction point, throughout the full year of 2012 [9], corresponding
to around $1.6 \cdot 10^{15}$ $pp$ collisions. The dominant Higgs production mechanism at
$\sqrt{s} = 8$ TeV is gluon fusion, a process characterized by a cross section of the order
of 20 pb. Therefore, in the full 2012 run, around $4.7 \cdot 10^5$ Higgs bosons were produced in the center of the CMS detector. Considering the branching fractions of
the experimentally most relevant decay modes and acceptance and efficiency limitations of the experiments, the total number of observed Higgs bosons is obviously
considerably smaller.

Besides the search for the Higgs boson and the analysis of its properties, there
are many other fields of research pursued at the LHC, not addressed here. Additional
information may be found in Refs. [2, 3] and in the several hundreds of physics
publications prepared by the LHC experiments. The physics agenda of the LHC
experiments, exploiting $pp$ collisions, can be split in two main categories. Firstly,
precision measurements of SM processes are conducted with the hope of finding
effects hinting at physics beyond the standard model (BSM), to better constrain SM
parameters, and to better understand the irreducible backgrounds of many LHC data
analyses. Examples for such studies are the searches for and measurements of the
rare $B_{s,d} \rightarrow \mu\mu$ decays, the measurement of CP violating phases in B-hadron decays,
measurements of the properties of the top-quark and of the W and Z bosons, as well
as the studies in quarkonium production physics presented in this thesis.

Secondly, direct searches for BSM physics are conducted exploiting a huge variety
of experimental signatures in the context of various theoretical models. Examples

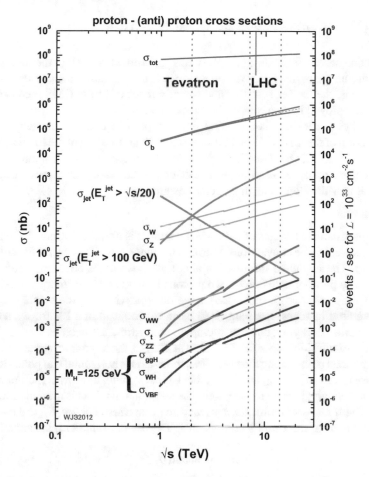

**Fig. 3.2** Standard model cross sections of various processes in $pp$ and $p\bar{p}$ collisions, as a function of $\sqrt{s}$ [10]

of such studies are searches for supersymmetric particles, searches for new massive vector bosons, direct dark matter searches, searches for extra dimensions, as well as searches for a $4^{th}$ generation of quarks and leptons. While the searches for new physics at the LHC have not yet uncovered any surprises, the LHC physics community has strong hopes that the $pp$ collisions at $\sqrt{s} = 13$ TeV starting in 2015 will lead to discoveries paving the way towards an improved theory, allowing for a unification of the gravitational interaction with the SM, as well as an improved understanding of dark matter.

## 3.2  The Compact Muon Solenoid Experiment

This section describes the design of the CMS experiment, as well as the running conditions during the periods used to collect the data used in the data analyses described in this thesis, the trigger systems, and specific features of the CMS reconstruction software employed in quarkonium physics analyses.

The CMS experiment is a general purpose particle detector located in Cessy, France, north of the main CERN facilities. It is built around one of the four interaction points of the LHC, the so-called "Point 5". The detector uses a combination of technologies employed in earlier particle detectors that have proven to be successful, as well as new developments in detector design.

**Conventions**

Before discussing the details of the CMS detector, it is useful to introduce the coordinate system and some conventions. During one bunch crossing, several PU collisions can occur. The complete information about all these collisions is denoted as one "event". The point of collision of two protons is denoted as "primary vertex" (PV), which is inside the "beam spot" (BS), a region approximating the overlap region of the two proton bunches. If a long-lived particle produced in a PV travels a certain distance before decaying, one can attempt to reconstruct a "secondary vertex" (SV). The center of the global coordinate system of CMS is the nominal "interaction point" (IP). The $z$-axis points along the beam line, the $x$-axis is horizontal and points towards the center of the LHC ring, and the vertical $y$-axis points upwards. The polar angle $\vartheta$ is measured with respect to the $z$-axis, the azimuthal angle $\varphi$ is measured from the $x$-axis. The transverse momentum $p_T$ and the transverse energy $E_T$ are measured from the values of the $x$ and $y$ components. The pseudo-rapidity $\eta$ is defined as [2]

$$\eta = -\ln\tan(\vartheta/2) . \tag{3.4}$$

## 3.2.1  Design

The design of the CMS experiment is driven by the aim to be able to address the LHC physics agenda in the best possible way, through the accurate reconstruction of all main experimental signatures of the processes of interest. Many signatures of interest involve muons, for which a good reconstruction efficiency, muon identification and momentum resolution is required. Furthermore, the design of the tracking system for the reconstruction of charged particles has to ensure a good reconstruction efficiency, momentum resolution, and a good identification of b-jets, which requires tracking detectors with very good spatial resolution close to the IP, in order to resolve SVs. The important features of the electromagnetic calorimeter (ECAL) include a very good diphoton and dielectron energy resolution, a large geometrical coverage, a good localization of the primary vertex and a handle on the direction of the reconstructed

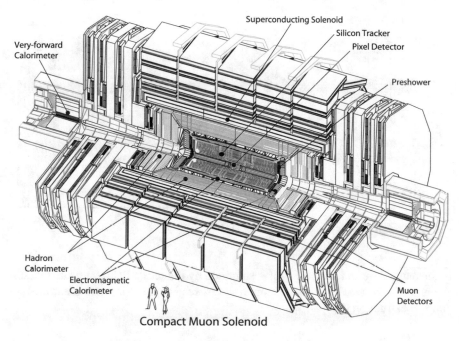

**Fig. 3.3** Cut-away view of the CMS detector [2]

photons. The hadron calorimeter (HCAL) is the main subdetector to measure the missing transverse energy, therefore requiring a large geometrical coverage and good resolution in $\eta$ and $\varphi$ [2].

**Detector Overview**

The CMS detector has a cylindrical symmetry along the $z$-axis, with the "barrel" at mid-rapidity closed by the two "endcaps" at forward-rapidity. The overall detector is 21.6 m long, has a diameter of 14.6 m and a weight of 12500 t, more than 95% of which is accounted for by the magnet system. Figure 3.3 shows a cut away view of the full detector, while Fig. 3.4 shows a transverse slice through the barrel, including sketches of the interaction of several groups of particles passing through and interacting with the subsystems of CMS. The individual subdetector systems are arranged in the overall design similarly to the layers of an onion. The innermost layer is the silicon tracker, used for the reconstruction of the track parameters of charged particles and the identification of PVs and SVs. The next layers are the crystal-based ECAL, which uses lead tungstate to produce scintillating light and measures the energy deposit of photons and electrons, and the HCAL, a brass/scintillator sampling calorimeter, mostly used to measure jet energies and the missing transverse energy. Beyond the calorimetry systems there is the decisive feature of the CMS detector, a 3.8 T superconducting solenoid, which is needed to bend the trajectories of charged particles, and to accurately determine their charges and momenta. The coil

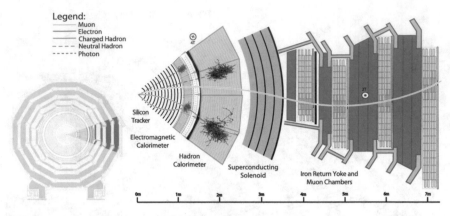

**Fig. 3.4** Transverse slice of the CMS detector [11]

is surrounded by an iron return yoke, saturated by the return field. Integrated within the return yoke are the various muon systems, responsible for muon identification and reconstruction [2].

### 3.2.2 Tracking Detectors

The innermost subdetector is the silicon tracking system. Its task is to efficiently and accurately track the trajectories of charged particles originating from the collisions. From the reconstructed tracks one can identify the PVs of the collisions, as well as the SVs of long-lived particles, which is of fundamental importance. Moreover, the particle momentum can be determined from the curvature of the tracks, caused by the Lorentz force induced by the particle momentum, its charge, and the practically homogeneous magnetic field within the volume of the tracker.

At the LHC nominal design, of the order of a thousand charged particles pass through the detector each bunch crossing, every 25 ns. This emphasizes the need for a fast, radiation-resistant tracker with high granularity. The choice of silicon detectors as CMS's tracking system is based on these requirements, ensuring efficient tracking up to $|\eta| = 2.5$ and for transverse momenta exceeding 1 GeV [4].

The high granularity in the high occupancy region close to the IP is achieved by employing silicon pixel detectors in the three innermost layers, surrounded by silicon strip detectors. In total, the CMS tracking system consists of approximately $6.6 \cdot 10^7$ silicon pixel detectors and $9.6 \cdot 10^6$ silicon strip detectors, covering in total an area of around $200 \, \text{m}^2$, making this by far the largest silicon detector ever built [4].

Signals from the pixel and strip detectors are clustered, combining signals from close-by detectors, building a "hit", characterized by a position and the corresponding uncertainty.

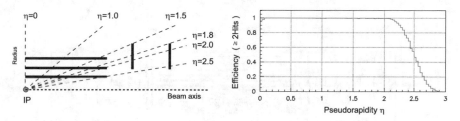

**Fig. 3.5** Layout of the silicon pixel detectors (*left*) and hit coverage as a function of |η| (*right*) [4]

### Silicon Pixel Detectors

The CMS pixel detector is composed of the barrel pixel detector (BPix) and the endcap pixel detector (FPix). The BPix consists of three layers of pixel modules with sensor cells of $100 \times 150$ μm², at radii $r = 4.4$, 7.3 and 10.2 cm. Each BPix layer is 53 cm long. The BPix is complemented by two endcap disk layers in the $x$-$y$ plane, at $|z| = 34.5$ and 46.5 cm, with radii between 6 and 15 cm. The geometrical layout of the pixel tracker is shown in Fig. 3.5, together with the hit coverage as a function of |η|. The arrangement of the BPix and FPix layers ensures that at a minimum of three pixel layers are hit in most of the region with |η| < 2.5. The spatial resolution of the BPix is of the order of 15–20 μm [4].

### Silicon Strip Detectors

The pixel detector is surrounded by the strip detector, where the particle flux density is smaller, and therefore less granular silicon strip detectors can be used. The geometrical layout of the strip detector is shown in Fig. 3.6. The barrel strip detector consists of a total of 10 layers, while the endcap strip detector is composed of 12 disks. The barrel part consists of the tracker inner barrel (TIB), close to the pixel detector, and the tracker outer barrel (TOB). The endcap part is composed of the tracker inner disk (TID) and the tracker endcap (TEC). At intermediate radii ($20 < r < 55$ cm, TIB and TID) the particle flux density allows the use of strip detectors with a typical size of 10 cm × 80 μm. At higher radii ($55 < r < 110$ cm), the size of the strip detectors can be further increased due to decreasing particle flux and occupancy, to typical strip sizes of 25 cm × 180 μm [4].

In some of the layers a second strip detector is added, as indicated in Fig. 3.6 by the double lines, back-to-back to the first strip detector and tilted by a small angle, resulting in so-called "stereo-hits", making it possible to measure the $z$-coordinate in the barrel and the $r$-coordinate in the endcap. The single-point resolution in the TIB is between 23 and 35 μm. The chosen layout of the strip tracker ensures that within the full range of |η| < 2.5 a charged particle passes at least nine strip detectors, at least four of them being stereo-modules [4].

The material budget of the silicon tracker is illustrated in Fig. 3.7 (left) as a function of $\eta$, represented by the number of interaction lengths, clearly showing an increase of material at forward rapidities.

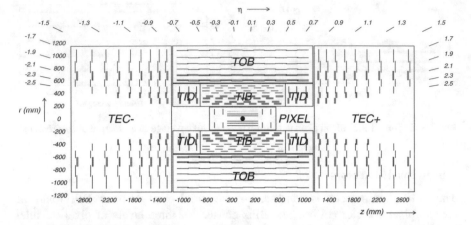

**Fig. 3.6**  Layout of the inner tracking system [4]

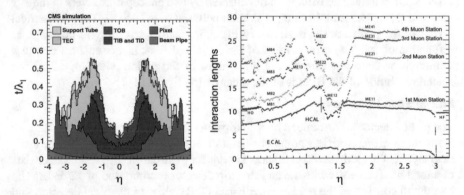

**Fig. 3.7**  Material thickness represented by the number of interaction lengths, as a function of $\eta$, for the tracker [12] (*left*) and up to various other subdetector systems [4] (*right*)

### 3.2.3  Muon Detectors

Efficient and accurate muon reconstruction is of fundamental importance for most of the experimental signatures studied by CMS, including those exploited for the analyses described in this thesis. The muon system fulfills three basic functions: identification of muons, measurement of their momenta, and triggering. The muon systems are the outermost detectors, integrated in the return yoke structures of the magnet system. One essential feature is that the other particles, except for very weakly interacting particles, such as neutrinos, are absorbed by the material they pass through before entering the muon detectors. Figure 3.7 (right) shows the material thickness in terms of numbers of interaction lengths up to the individual elements of the muon system, as a function of $|\eta|$.

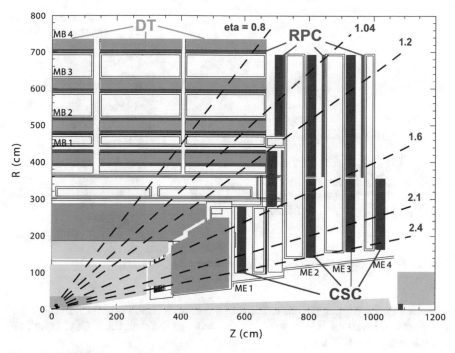

**Fig. 3.8** Layout of the muon system [4]

The geometrical layout of the muon detectors is shown in Fig. 3.8. The muon detectors consist again of a cylindrical barrel region and two endcap disks. The barrel region is made of five "wheels". The endcap disks are divided in two concentric rings, except for the first one, which is split in three concentric rings. Due to the large area covered by active detection planes of around 25 000 m², an inexpensive but robust solution had to be found. The system is composed of three different types of gaseous detectors [4].

In the barrel region, drift tubes (DTs) are used, and are organized in four muon stations (MB1–MB4), placed in between the layers of the iron return yoke. The stations MB1–MB3 consist of eight chambers each, four measuring the $r\varphi$ coordinates, and four measuring the $z$ component. The station MB4 only contains the four chambers measuring $r\varphi$. To avoid dead-spots, the drift cells are offset by a half-cell width with respect to the neighboring chambers. The barrel region is characterized by low muon rates and a uniform magnetic field [4].

In the endcaps, where the muon rate is large and the magnetic field is non-uniform, the muon detectors consist of cathode strip chambers (CSCs). They are organized in four muon stations of CSCs (ME1–ME4) in between the flux return plates, positioned perpendicular to the beam line, measuring the coordinates in the $r\varphi$ plane [4].

In addition to the DTs and CSCs, in both barrel and endcaps, resistive plate chambers (RPCs) are also used. These detectors have a worse spatial resolution than the DTs and CSCs, but a very fast response and excellent time resolution, which is

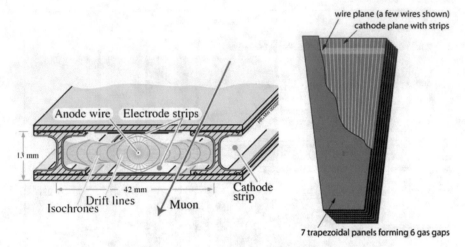

**Fig. 3.9** Sketches of DT (*left*) and CSC (*right*) chambers [4]

exploited by the trigger system to identify the correct bunch crossing. In the barrel there are six layers of RPCs, while in the endcaps the first three CSC layers are complemented by RPCs [4].

**Drift Tube Chambers**

The DT chambers are filled with a mixture of argon (85%) and $CO_2$ (15%). If a muon enters the chamber, gas molecules are ionized and the resulting ions and electrons drift to the respective electrodes. The time the muon enters the chamber is precisely determined by the neighboring RPCs. From the measured drift-time and the known drift-velocity, one can calculate the coordinates of the muon trajectory through the chamber, with a resolution of around 1 mrad in direction, and around $100\,\mu m$ in position [4]. Figure 3.9 (left) shows a sketch of a DT chamber.

**Cathode Strip Chambers**

CSCs are multiwire proportional chambers in trapezoidal shape, as sketched in Fig. 3.9 (right), consisting of 6 gas detector layers (40% argon, 50% $CO_2$, 10% $CF_4$) filled with an anode wire plane each, and interleaved with seven cathode panels. If a muon enters the CSC, gas molecules are ionized, inducing a fast signal in the anode wires, which is used in the muon trigger system. The cathode signal is slower, but allows to reach a better spatial resolution of around 10 mrad in direction, and around $200\,\mu m$ in position [4].

**Resistive Plate Chambers**

RPCs are gaseous parallel-plate detectors, filled with a mixture of $C_2H_2F_4$ (96.2%), $C_4H_{10}$ (3.5%), and $SF_6$ (0.3%). They are operated in avalanche mode and have a time resolution of the order of 1 ns [4].

### 3.2.4 Trigger and Data Acquisition Systems

This section describes the CMS trigger and data acquisition (DAQ) systems, which are the most crucial parts of the CMS data taking scheme. The LHC delivers billions of collisions per second to CMS, with a bunch crossing rate of 40 MHz at nominal design. This amount of data is simply too much to be processed and stored, mostly because of limitations in the capabilities of the online computer farm. Fortunately, the majority of the $pp$ collisions do not contain interesting physics and can be rejected. The selection of the "interesting" events is done with the CMS trigger system. The output of CMS is limited to the order of $10^2$ Hz. Therefore, a rejection factor of around $10^6$ has to be achieved. The CMS trigger system is divided in two steps, a hardware-based "level 1 trigger" (L1) and a software-based "high level trigger" (HLT), running on an online computer farm, further reducing the number of events to be stored. Simultaneously, the performance of the individual subdetector systems has to be monitored continuously, and a fraction of the selected events is routed to online services which perform "data quality monitoring" (DQM).

The accepted events are then forwarded to mass storage devices and their content is reconstructed, identifying high-level analysis objects, before being distributed to a world-wide system of computing and storage facilities, where they can be accessed to perform high-level physics data analysis [4].

**Running Conditions in the LHC Run I**

The load on the DAQ, trigger, event reconstruction and storage systems heavily depends on the running conditions provided by the LHC. These conditions have changed dramatically throughout LHC Run I, as can be appreciated in Figs. 3.10 and 3.11. The top panel of Fig. 3.10 shows the peak instantaneous luminosity as a function of time throughout the full Run I operation. While the maximum peak instantaneous luminosity in 2010 was around $2 \cdot 10^{32}$ cm$^{-2}$s$^{-1}$, it increased in 2011 to around $4 \cdot 10^{33}$ cm$^{-2}$s$^{-1}$ and to $7.7 \cdot 10^{33}$ cm$^{-2}$s$^{-1}$ in 2012, close to the nominal LHC design luminosity of $10^{34}$ cm$^{-2}$s$^{-1}$, planned to be reached in LHC Run II. The bottom panel of Fig. 3.10 shows the cumulative integrated luminosity, as delivered by the LHC, as a function of time. The slope of the cumulative luminosity increases proportionally to the instantaneous luminosity. The integrated luminosity delivered to CMS was 44.2 pb$^{-1}$ in 2010, 6.1 fb$^{-1}$ in 2011, and 23.3 fb$^{-1}$ in 2012 [9].

With increasing instantaneous luminosity, the number of $pp$ collisions per bunch crossing, i.e., the number of PU collisions, also increases, as can be appreciated in Fig. 3.11, which shows the peak number of interactions per crossing for each day of running in LHC Run I. The average number of collisions per crossing in 2012 was 21, with tails extending up to around 40 [9]. The PU is a challenging aspect that can introduce difficulties in the data analyses. The tracking algorithms are affected by the track multiplicity originating from several PVs. The individual PVs have to be identified carefully, not to bias the analyses. However, the PU has a negligible effect on the quarkonium physics analyses discussed in this thesis, given

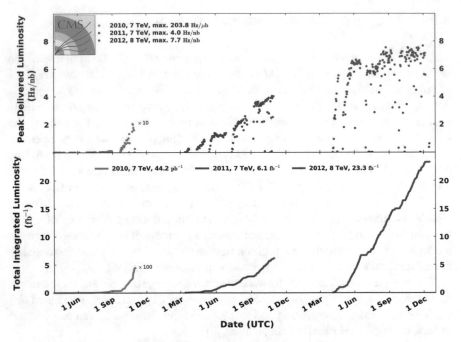

**Fig. 3.10** Peak instantaneous luminosity (*top*) and cumulative integrated luminosity (*bottom*), as delivered by the LHC to CMS, as a function of time throughout the full Run I [9]

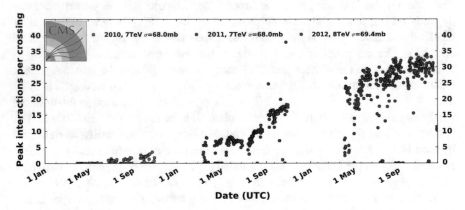

**Fig. 3.11** Peak number of *pp* collisions per bunch crossing, as delivered by the LHC to CMS, as a function of time throughout the full Run I [9]

that the experimental signature contains two muons forming a clear PV that is almost unambiguously identifiable.

The full CMS efficiency, the fraction of the collisions provided by the LHC that could be efficiently processed and recorded, is very high, ranging from 90.5 to 93.5%, slightly different for the different running periods [9].

**Fig. 3.12** Overview of the L1 trigger [13]

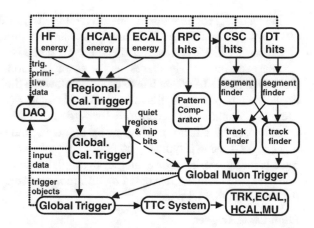

## Level 1 Trigger

The target of the L1 trigger step is to reduce the L1 output rate to less than 100 kHz. The decision regarding if an event is accepted or not by the L1 trigger can only be delayed for 3.2 μs after the bunch crossing, including the transfer of the trigger information to the L1 facilities. During this time the data is conserved in buffers. These timing requirements limit the information used in the L1 to promptly available data, excluding information from the silicon tracker. In addition, no expensive processing can be conducted by the L1 trigger calculations. Therefore, only information from the muon systems and calorimetry is used for the L1 decision. If the L1 decision is positive, the data is moved to a buffer to be read out and further processed by the HLT [9].

The L1 hardware consists of programmable custom electronics. Figure 3.12 shows a diagram of the L1 layout, with all its major subcomponents. The L1 consists of local, regional and global components. The local triggers, so-called "trigger primitives", are based on hits in the individual muon detectors (DTs, CSCs and RPCs), and on energy deposits in the calorimeters. The regional triggers combine the information from the trigger-primitives of a given spatial region, using pattern logic to identify trigger objects such as muons or electrons. The global muon and calorimeter triggers combine the information of the full respective subsystem, forwarding the information to the L1 global trigger, which is the highest-level entity in the L1 trigger system. Finally, the L1 global trigger can make a decision to accept or reject a given event, based on the information provided by the global muon and global calorimeter triggers [9].

The L1 triggers are limited to calculate simple quantities of the L1 objects. As an example, for muon objects the L1 is able to identify with a certain precision the muon charge, $\eta$ and $p_T$. The muon $\eta$ is not determined in a continuous way, but determined in discrete steps, so-called "$\eta$ L1 indices". The L1 muon $p_T$ resolution is about 15% in the barrel and 25% in the endcaps [4].

Individual L1 triggers, also denoted as "L1 seeds", are developed, requiring certain physics objects, subjected to certain cuts. They are grouped together in a so-called "L1 menu", which evolves over time. The L1 menu is composed of seeds covering various experimental signatures, as for example muons and jets. The L1-accepted rate of the whole L1 menu has to be kept below the limit of 100 kHz, which can be done by either prescaling the individual seeds by $n_{PS}$ (meaning that only every $n_{PS}^{th}$ event that would be L1-accepted is forwarded to the HLT, reducing the rate of this specific seed by a factor of $n_{PS}$) or by tightening the requirements of the physics objects. A L1 menu is typically composed of the order of 100 different seeds, whose rates partially overlap. For example, events L1-accepted by seeds only requesting one muon are often also accepted by different seeds requesting two muons, sharing the corresponding L1 rate.

**High Level Trigger**

The limit of the final event rate is of the order of a few hundred events per second. This limit has increased throughout the LHC Run I period, exceeding the nominal design of the CMS trigger system. The HLT system takes as input all events that are accepted and forwarded by the L1, and performs more sophisticated software reconstruction algorithms on a dedicated processor farm, using commercial CPUs, using information from all subdetector systems, including the tracker, in order to make a more refined decision about the usefulness of the event, compared to L1. During the reconstruction, the information content of the full event has to be stored in buffers, given that the HLT algorithms can take up to a few seconds for the decision to accept or reject an event, especially for high-PU events [9].

There are a few hundred HLT algorithms. These algorithms are denoted as "HLT paths", and the combination of all paths is denoted as the "HLT menu". As in the case of the L1 menu, the HLT menu evolves over time, depending on the instantaneous luminosity of the LHC. Each HLT path starts from a specific L1 seed (or a logical combination of several seeds), running only on the events that were L1-accepted by the specific seed(s). The HLT algorithms are organized as an alternate sequence of "HLT producers", which perform part of the event reconstruction, and "HLT filters", which can reject the event based on the reconstructed quantities of the producer. In this way, one can optimize the CPU time needed for each HLT path by first running the parts of the reconstruction which are simple and fast, and possibly already rejecting the events at an early stage, before the expensive reconstruction steps such as tracking algorithms start to be executed. Only if the last filter of a path is passed, the event is accepted, and subsequently saved for offline physics analysis.

The HLT online reconstruction algorithms are designed differently than the offline ones, as the time per event needed for the HLT reconstruction is a vital parameter of the system that needs to be minimized. Therefore, the objects at HLT are affected by worse resolutions than the corresponding objects at the offline reconstruction level.

The individual HLT paths are grouped according to the different physics analysis groups (PAGs). The accepted events are streamed into the corresponding primary data sets (PD), e.g., the "MuOnia" PD for the b-physics (BPH) PAG, which is responsible for the development, maintenance and validation of the quarkonium physics triggers.

The sum of the rates of all PDs has to be within the limit of the overall HLT output rate. The relative fraction of the total bandwidth allocated to the individual physics groups is decided by the CMS physics management, reflecting the priorities of the physics agenda of the experiment.

One of the main limitations of the HLT output rate is the "prompt reconstruction" of the HLT-accepted events, which is typically done up to 48 h after the data are taken. In 2012, the HLT output rate was roughly doubled thanks to the technique of "data parking". Given that the 2012 run was followed by the LHC shutdown period, freeing huge amounts of computing resources otherwise busy with prompt reconstruction, it was feasible to define a set of "parked" HLT paths, from which events were collected online and put into storage without reconstruction. The full reconstruction of these events was then performed after the 2012 data taking period, during the 2013 shutdown period, when resources became available. This technique was especially advantageous for the triggers used for quarkonium physics.

**Dimuon Triggers**

All the analyses described in this document rely on triggers that request two muons at both the L1 and the HLT, so-called "dimuon triggers". The basic characteristics of these triggers are discussed here. Above a certain instantaneous luminosity, only requesting the presence of a dimuon with no other requirements does not fit into the allocated bandwidth. Possible handles to reduce the dimuon rate at L1 include requirements on the dimuon kinematics $p_T$ and $y$, on the muon $p_T$ and $\eta$, on the $\Delta\eta$ of the two muons, as well as a muon quality requirement (for example "high quality"), as well as the requirement that the two muons have opposite charges. The latter requirement was not used in LHC Run I, but will be employed in the higher luminosity environment of LHC Run II. At the HLT, one can use the same handles as at L1, plus additional ones, thanks to the more sophisticated reconstruction and to the information from the tracker. These extra handles include cuts on the dimuon invariant mass, cuts on the distance of closest approach (DCA), the $\chi^2$ probability of the dimuon vertex fit ($P_{vtx}^{\mu\mu}$) and cuts on the transverse displacement of the dimuon vertex, the latter one not used for quarkonium physics triggers.

The changes in the LHC running conditions and the HLT bandwidth limitations during the Run I period are reflected in the dimuon L1 and HLT strategies, as can be appreciated in Fig. 3.13, showing overview plots of the 2010 and 2011 dimuon triggers, and of the 2012 parked dimuon triggers. The quarkonium dimuon triggers are highly efficient, and collect only a small fraction of background events. In order to reduce the rate, one necessarily has to reject signal events. This could be done with prescales, but it is more reasonable to reject low-$p_T$ quarkonia to prioritize the analysis of the high-$p_T$ region of quarkonium production, considering the physics motivation of these measurements (see Chaps. 2 and 5).

In 2010 the dimuon rate was low enough such that one simple trigger could be used, covering all quarkonium mass regions, without the need to apply kinematical constraints on the muons or dimuons, neither at L1 (*L1_DoubleMu0*) nor at HLT (*HLT_Dimuon0*).

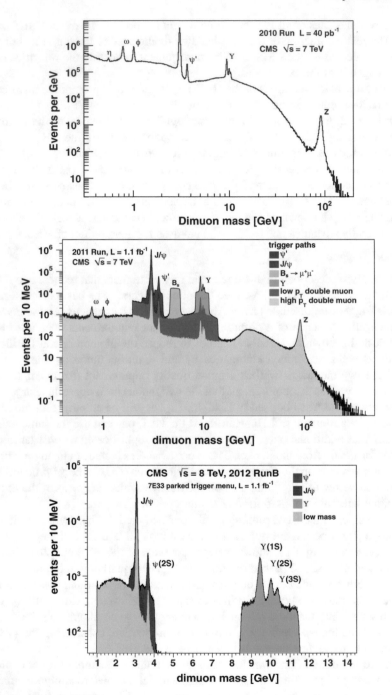

**Fig. 3.13** Overview of the dimuon triggers as used in the 2010 [14] (*top*) and 2011 [15] (*middle*) data taking periods, as well as of the parked dimuon triggers in 2012 (*bottom*)

In 2011 the situation changed, due to the increasing instantaneous luminosities. While the L1 dimuon seed was only complemented by a high-quality requirement for both muons ($L1\_DoubleMu0\_HighQ$), the dimuon HLT paths used for quarkonium physics changed dramatically with respect to the one used in 2010, in order to reduce the corresponding rates. The quarkonium triggers were split into three mass regions for the $J/\psi$, $\psi(2S)$ and $\Upsilon(nS)$ analyses. A minimum dimuon $p_T$ was imposed, at values which were adapted throughout the year, depending on the instantaneous luminosity. The dimuon rapidity was restricted to the barrel region, $|y| < 1.25$ (except for the $\psi(2S)$ trigger), and so-called "cowboy dimuons", which are geometrically aligned in a way such that the magnetic field bends the two muons towards each other, $\varphi(\mu^-) - \varphi(\mu^+) < 0$, were rejected. Furthermore, a DCA cut and a cut on $P_{vtx}^{\mu\mu}$ were added. The HLT paths can be identified as $HLT\_DimuonX\_Jpsi\_Barrel$ (with X = 10 or 13, corresponding to the changing minimum dimuon $p_T$ requirement), $HLT\_DimuonX\_PsiPrime$ (with X = 7, 9 or 11) and $HLT\_DimuonX\_Upsilon\_Barrel$ (with X = 5, 7 or 9). The shift of strategy from inclusive to more exclusive triggers can be clearly seen in the top two panels of Fig. 3.13.

For the 2012 data taking period, the quarkonium L1 seed remained almost unchanged ($L1\_DoubleMu0er\_HighQ$), except for the loose requirement that both muons be within $|\eta| < 2.1$. Due to the possibility of data parking, the requirements of the quarkonium HLT paths could even be loosened with respect to 2011, by lowering the dimuon $p_1$ thresholds and removing the barrel-requirement and the cowboy-rejection. The corresponding HLT paths can be identified as $HLT\_DimuonX\_Jpsi$ (with X = 8, 10), $HLT\_DimuonX\_PsiPrime$ (with X = 5, 7) and $HLT\_DimuonX\_Upsilon$ (with X = 5, 7).

The L1 and HLT trigger rates are proportional to the instantaneous luminosity. At the beginning of a LHC fill, when the $pp$ beams are first brought to collision, the instantaneous luminosity and the trigger rates are highest, then decreasing with time, as the $pp$ beams lose intensity and focus. This effect can be seen in Fig. 3.14, showing the quarkonium L1 and HLT rates for a typical LHC run in 2012. The L1 and HLT menus have to be designed such that the peak rates at the beginning of the LHC fills are below given thresholds.

Besides these dimuon triggers used for physics analyses, also so-called "efficiency triggers" are very important, running prescaled at very low rates, collecting events with looser requirements on the muons. These triggers are used to estimate the muon efficiencies with data driven methods.

**Event Reconstruction and Processing**

Once an event passes the L1 and HLT requirements, it is saved and reconstructed. The full information of the CMS subdetectors is transferred to the so-called Tier-0 (T0) center at CERN, saved in the RAW data format, with a size of around 1.5 MB per event, and stored permanently. There, the events are subjected to the full offline reconstruction algorithms, employing the most sophisticated approaches used in the reconstruction chain, then saved in the RECO format, with a size of around 0.25 MB per event, which contains already the high-level physics analysis objects. The CMS data are saved in the ROOT format [17] and the reconstruction is conducted by the

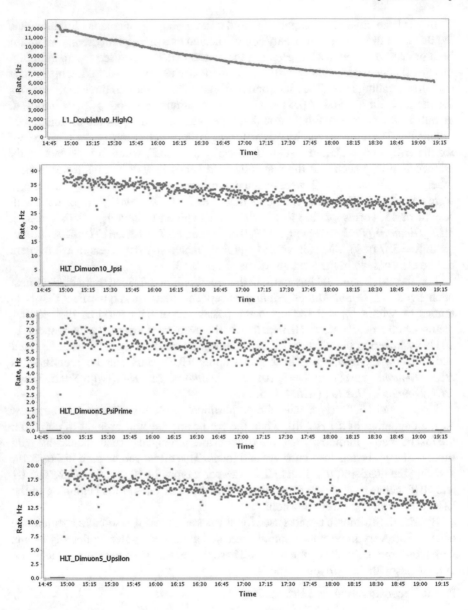

**Fig. 3.14** Dimuon trigger rates for a typical *pp* run in 2012 at an instantaneous luminosity of around $7 \cdot 10^{33}$ cm$^{-2}$s$^{-1}$ at the beginning of the LHC fill, of the quarkonium L1 seed (*top*), the J/$\psi$ (*second row*), $\psi(2S)$ (*third row*) and $\Upsilon(nS)$ (*bottom*) HLT paths [16]

CMS software (CMSSW) framework, which is based on object oriented C++ classes steered by python configuration files. The following Sect. 3.2.5 introduces the parts of the full offline reconstruction software that are most relevant for this thesis.

The T0 facility distributes the RAW and RECO datasets to one of the seven Tier-1 (T1) centers that exist around the world. There, the datasets are stored and can be re-reconstructed in case this is needed, due to changed calibration constants or improvements in the reconstruction software. The re-reconstruction at the T1 centers can be limited to a subset of the RECO event content, resulting in datasets in so-called AOD (analysis object data) data format, which is smaller in size than the RECO format, with a size of around 0.05 MB per event, but contains sufficient information for most physics analyses. The T1 centers distribute the RECO and AOD data to the many Tier-2 (T2) centers around the world, such as for example the T2 center operated by the Institute of High Energy Physics (HEPHY) in Vienna. This is the location where finally the high-level physics analyses can be conducted, accessing the files at the storage facilities of the T2 centers, and analyzing them with their computing facilities, usually producing n-tuples of the events, containing all information needed for the physics analyses, which can be saved at any other T2 [4]. Besides the data collected at CMS, many analyses rely on Monte Carlo (MC) simulation samples, which are also produced and distributed through the Tier-system. The size of the RAW data format per event for simulated data is around 2 MB, due to additional MC truth information.

### 3.2.5   Offline Track and Muon Reconstruction

Track and muon reconstruction are very important and complex features of the CMS reconstruction software. Track objects are used for the reconstruction of muons, converted photons, electrons, taus, charged hadrons and jets, and are therefore vital objects for most of the experimental signatures studied at CMS. As quarkonium physics analyses rely on muon and photon conversion reconstruction, it is imperative to summarize the techniques used for these reconstruction steps.

**Track Reconstruction**

In a homogeneous magnetic field, one can describe the trajectory of a charged particle by a helix, parametrized by five helix parameters, also denoted as track parameters. These depend on the particle charge, momentum, and a given starting point that can be chosen arbitrarily and is usually taken to be one hit of the given track. The actual trajectory depends on inhomogeneities of the magnetic field, multiple scattering and energy loss while interacting with the detector material. These effects have to be taken into account carefully in the track reconstruction.[3] The first step of the CMS tracking is the definition of the BS and the corresponding uncertainties, which are around $\sigma_x$, $\sigma_y = 15$–$20$ $\mu$m and $\sigma_z = 5$ cm. The BS is taken as the initial estimate of the primary interaction point, in the transverse plane. Starting from the BS, the

---

[3]These effects are taken into account in all mentioned extrapolations of track trajectories in between detector layers or in between subdetector systems, even though not mentioned explicitly in each occurrence.

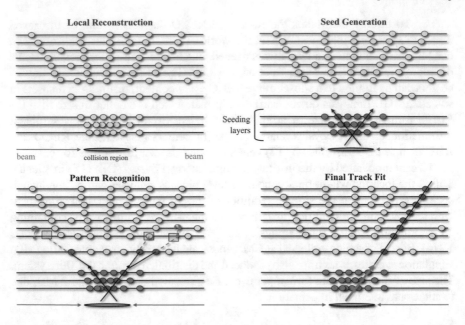

**Fig. 3.15** Sketch of the CMS track reconstruction: local reconstruction (*top left*), seed generation (*top right*), pattern recognition (*bottom left*) and the final track fit (*bottom right*) [19]

first stage of the track and vertex reconstruction is initiated, using only information from the pixel detectors. These initial vertices are then used in the full CMS tracking, which uses the combinatorial track finder (CTF), starting from the hits reconstructed in the silicon tracker [18].

The track reconstruction is performed in an iterative way, repeating the CTF several times. The individual iterations are characterized by different seeding algorithms and different requirements on the track quality and kinematics. They aim at the reconstruction of specific objects, different "categories" of tracks, such as high-$p_T$ and low-$p_T$ prompt tracks in the first iterations, and the reconstruction of tracks from displaced vertices in later iterations. After each iteration, the fully reconstructed tracks are added to the list of tracks from the previous iterations and the reconstructed hits forming these reconstructed tracks are removed, to reduce the combinatorics of the following iterations. In this way, the tracks of the individual categories can be reconstructed efficiently, while ensuring that the algorithms make optimal use of the available computing facilities. The reconstruction of displaced tracks, for example, would not be feasible if attempted starting from all reconstructed hits ("local reconstruction") [18]. The CTF, used in each iteration of the track reconstruction, is split in four main parts, the seed generation, pattern recognition, ambiguity resolution and the final track fit. Figure 3.15 shows a sketch of the individual parts of the reconstruction.

**Seed generation**: So-called "trajectory seeds" are the starting point for the pattern recognition algorithms. In the ideal case, the seed constrains all of the five track parameters in an unbiased way and with sufficiently small uncertainties. Given that the momentum of a trajectory can only be estimated if at least three trajectory positions are known, the minimal set of objects to build a trajectory seed are either two hits plus a vertex constraint, or three hits from different layers. The main tracking iterations use three hits, so-called "pixel triplets" for the seeding. The efficiency of pixel triplet seeding is significantly smaller than the efficiency of hit pair plus vertex seeding, but the purity of the triplet seeds is higher. In later iterations, also mixed seeds from pixel and strip detector hits are used, as well as strip-only triplet seeds, mostly for the reconstruction of displaced tracks [20]. Besides the momentum and position information of the trajectory seed, so-called "tracking regions" are defined, and added to the seed object. The tracking regions define the limits of the acceptable track parameters, and are used in the pattern recognition step, reducing the corresponding combinatorics [12].

**Pattern recognition**: Starting from each seed, the pattern recognition algorithms first determine which layers are compatible with the seed trajectory. Using a combinatorial Kalman filter method [21], the trajectory is extrapolated to these compatible layers, followed by a search, within these layers, for hits compatible with the extrapolated trajectory. If several compatible hits are found, a new trajectory candidate for each hit is built. Even if no hit is found, a virtual hit is created, to take into account the possibility that the charged particle did not leave a hit in this layer. The trajectory candidates are then updated with the information from the additional hit and its uncertainties. The procedure is iteratively continued outwards, until the outermost layer of the tracker is reached, or after a predefined number of consecutive layers without compatible hits have been crossed [22].

**Ambiguity resolution**: The pattern recognition algorithms create ambiguities, in cases where several trajectory candidates are built from the same seed, or in cases where the same track may be reconstructed from different seeds. In order to avoid double counting of tracks, these ambiguities have to be resolved, which is done based on the fraction of shared hits between two trajectory candidates [22].

**Final track fit**: This step starts from the trajectory candidates of the pattern recognition that pass the ambiguity resolution requirements and estimates the final track parameters. The trajectory candidates, with all associated hits, are refitted with a Kalman filter approach, from the innermost layer of the trajectory to the outermost layer. In a second step, the obtained trajectory is smoothed by employing a second Kalman filter, initialized by the information of the first filter in the outermost layer, and run backwards towards the innermost pixel layers. This procedure provides optimal estimates of the track parameters, especially on the first and last layers of the trajectory [22].

After each CTF iteration, tracks are subjected to certain quality requirements, to avoid the usage of fake tracks or tracks which are poorly reconstructed, with large uncertainties on the track parameters. These requirements include cuts on the track $p_T$ and $\eta$, and on the reduced $\chi^2$ of the track fit, on the number of hits in the pixel and strip layers, on transverse and longitudinal impact parameters and on their uncertainties. Several categories of tracks are defined, corresponding to different sets of requirements. The tracks failing the loosest requirements are rejected, the tracks fulfilling the tightest requirements are categorized as "high-purity" [23].

**Vertex Reconstruction**

Given the number of PU interactions provided by the LHC, the unambiguous determination of the PV is very important. The PV reconstruction algorithm is split in two parts, the vertex finding and the vertex fitting. Since the BS constrains all PVs to a narrow region in the transverse plane, vertex finding only operates along the $z$-axis. The tracks found by the iterative CTF procedure are clustered depending on their $z$ coordinates at the point of closest approach to the beam axis. These vertex candidates are then fit by an adaptive vertex fitter, using the corresponding track-cluster, in order to estimate the coordinates and uncertainties of the PV [23].

The resolution of the PV reconstruction depends on the number of associated tracks and their $p_T$, the resolution improving with higher-$p_T$ tracks and the number of tracks. The values of the PV resolution vary from around $10\,\mu m$ up to around $100\,\mu m$ for the transverse coordinates, and from $15\,\mu m$ up to around $150\,\mu m$ for the $z$ coordinate [12].

Besides the estimation of the signal PV, the reconstruction of the SVs is of importance, especially for the analyses described in this thesis. The SV reconstruction is conducted with a kinematic vertex fitting approach, using all tracks and vertices of the current decay and applying physics constraints reflecting the assumptions made about the specific decay process. The typical resolution of the SV coordinates is of the order of $30\,\mu m$, for two-body decays [10, 24].

**Tracking Performance**

The CMS silicon tracker is performing very well. The average hit efficiency in the pixel and strip layers is above 99%. The general track reconstruction is also performing very well, as can be appreciated in Fig. 3.16, showing the CTF tracking reconstruction efficiency, including algorithmic and hit efficiency as well as acceptance effects, of muons and electrons from a MC simulation. These results were obtained using seeds from the pixel detector only. The lower efficiency of the electrons is caused by interactions with the detector material, as for example the electron producing an electromagnetic shower within the tracker volume. This effect is more important in the endcaps than in the barrel due to the larger material budget (see Fig. 3.7). Electrons for use in physics analyses are reconstructed with a dedicated procedure, improving efficiency and resolution [12].

**Muon Reconstruction**

In the standard CMS muon reconstruction, tracks are first reconstructed independently in the inner silicon tracker, defined as "tracker track", and in the muon system, defined as "standalone muon track". Based on these two objects, there are two different muon reconstruction algorithms resulting in so-called "global muons" and

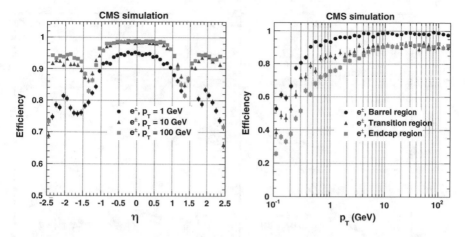

**Fig. 3.16** Tracking reconstruction efficiency of muons (*left*) and electrons (*right*), estimated from simulation, as a function of $p_T$, in different regions of $|\eta|$ [12]

"tracker muons", which are discussed in detail in Ref. [25], and briefly introduced here.

**Global muon reconstruction**: This is an "outside-in" method. Each standalone muon track is extrapolated back to the outermost layer of the silicon tracker, where a compatible tracker track is found by comparing the track parameters of the two tracks. The global muon track is built by refitting the combined tracker and standalone muon track, using a Kalman filter.

**Tracker muon reconstruction**: This is an "inside-out" method. Tracker tracks satisfying certain minimum momentum requirements are considered as possible muon candidates. Their trajectories are extrapolated outwards to the first muon station. If the algorithm finds at least one muon segment (a track stub consisting of DT or CSC hits) that is compatible with the extrapolated tracker track, it is labelled as tracker muon, without being updated by the information from the matched muon segment.

Figure 3.17 shows the transverse momentum resolution as a function of the muon $p_T$, when using only information from the tracker, the muon stations, or both systems. Below a $p_T$ of around 200 GeV, the momentum resolution is completely dominated by the information from the silicon tracker. Only for higher-momentum muons the additional information from the muon stations improves the momentum resolution [2].

The tracker muon reconstruction algorithm is rather CPU-intensive, due to the large number of tracker tracks that have to be extrapolated to the muon stations. It further leads to a non-negligible fraction of fake muon tracks, given that tracker muons only require one muon segment, while the standalone muon reconstruction requires, typically, at least two active muon segments. Therefore, global muons have a higher purity than tracker muons. Nonetheless, despite the drawbacks, tracker muons are more efficient than global muons, especially at low $p_T$, a relevant kinematic region for the analyses described in this thesis, which therefore use both tracker and global muons.

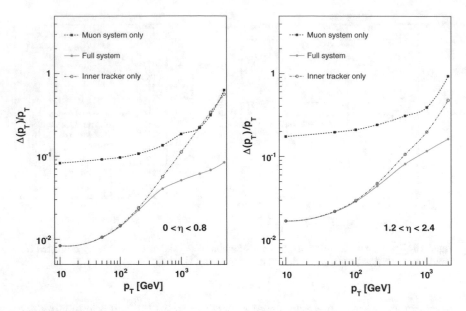

**Fig. 3.17** Muon $p_T$ resolution as a function of $p_T$, using the information from the tracker, from the muon stations, or from both, in the barrel (*left*) and endcaps (*right*) [2]

Due to the high reconstruction efficiency of the silicon tracker for muon tracks, as shown in Fig. 3.16, and the high efficiency of the muon detectors, muons with sufficiently high momentum are reconstructed by CMS with an overall efficiency of more than 99%, either as global or tracker muon, or as both [25]. In order to increase the purity of the reconstructed muons, muon identification criteria are applied. There are several sets of requirements, with different trade-offs between purity and efficiency, depending on the needs of the specific analysis. The specific sets of cuts, not discussed here, can be found in Ref. [25].

### 3.2.6 Photon Conversion Reconstruction

For S-wave quarkonium production analyses at CMS, the dimuon decay is exploited, for which two muons with opposite sign are combined to form a dimuon vertex. In order to reconstruct the radiative decays of the P-wave $\chi$ states, $\mathcal{Q}^{3P_J} \rightarrow \mathcal{Q}^{3S_1} + \gamma$, additionally the photon $\gamma$ has to reconstructed. Figure 3.18 shows example distributions from CMS 8 TeV data in the $J/\psi + \gamma$ channel. The top panel shows the mass distribution of the $\chi_c$ candidates, $M^{\chi}$, including an unbinned maximum likelihood (ML) fit of a suitable composite model describing all relevant signal and background contributions. The $J=1$ and $J=2$ states are rather close in mass, separated by only 45.5 MeV in the $\chi_c$, 19.4 MeV in the $\chi_b(1P)$ and 13.5 MeV in the $\chi_b(2P)$ systems

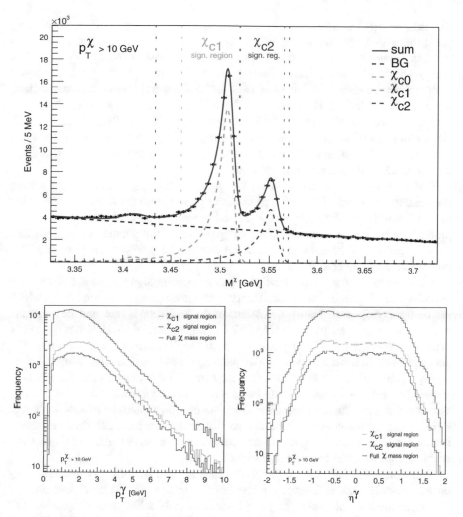

**Fig. 3.18** Invariant $\mu\mu\gamma$ mass distribution of the $\chi_c$ candidates (*top*), including a fit, and the reconstructed photon $p_T$ (*bottom left*) and $\eta$ (*bottom right*) distributions, for $\chi_c$ candidates with a $p_T > 10\,\text{GeV}$

[26]. A good energy resolution of the photon, leading to a good mass resolution of the $\mu\mu\gamma$ system, is required in order to separate the J=1 and J=2 states. The bottom panels of Fig. 3.18 show the reconstructed photon $p_T$ and $\eta$ distributions, for $\chi_c$ candidates with $p_T > 10\,\text{GeV}$ and $|y| < 1.2$. The $p_T$ distribution is rather soft, the bulk of the events being below 2 GeV, and the $\eta$ distribution is dominantly populated at mid-rapidity, below $|\eta| \approx 1$.

There are, in principle, two ways to reconstruct photons in CMS, either from the energy deposits in the ECAL, or through photon conversion in the silicon tracker. In order to resolve the J=1 and J=2 states, a relative photon energy resolution of the order of $10^{-3}$ is necessary. Photon reconstruction in the ECAL is excellently performing for large photon energies typical for the $H \rightarrow \gamma\gamma$ decays, as the relative calorimeter energy resolution improves with increasing energy. Given the low energies of the photons emitted in the radiative decays, quarkonium P-wave analyses with the requirement to resolve the $J = 1$ and $J = 2$ states cannot rely on photons reconstructed in the ECAL system [2]. Therefore, these analyses rely on the reconstruction of photon conversions, fully reconstructed with the silicon tracker, which allows for mass resolutions of the $\chi$ states sufficiently good to resolve the $J = 1$ and $J = 2$ states. This section describes the general conversion reconstruction algorithms used in CMS, and possible improvements.

Photon conversions occur through $e^{+}e^{-}$ pair production processes, which are the dominating processes of the interaction of the photon with the nuclei of the tracker material. For the photon energies considered in the P-wave analyses the conversion probability is, in first approximation, independent of the photon energy [26, 27], only depending on the amount of material that the trajectory of the photon passes through (see left panel of Fig. 3.7).

Photon conversions are characterized by two opposite-sign electrons, also denoted as dielectron, originating from a secondary vertex, the conversion vertex (CV), which can be highly displaced. The invariant mass of the dielectron is zero, and the momenta of the two electrons are parallel at the CV. The trajectories of the electrons open only in the transverse plane, affecting the angle $\varphi$, due to the magnetic field geometry. The photon momentum is not shared in equal parts by the two electrons but allows also for highly asymmetric momenta of the two electrons. For this reason, and given that the photons interesting for quarkonium P-wave analyses are generally characterized by very low energies, the conversion electrons are also characterized by low momenta. Below a certain momentum, electrons usually do not reach the calorimeter system, because they are absorbed by the tracker material and/or because the track helix stays within the tracker, due to the high magnetic field.

Therefore, the chosen strategy for the analysis of P-wave quarkonia is to rely on tracker-only information for the low-energy photon conversion reconstruction. The reconstruction of low-$p_{\mathrm{T}}$ displaced tracks heavily relies on the capabilities of the iterative tracking approach, as described above. Once the iterative tracking step is finished, the conversion finding algorithm is initiated. From the full list of tracks reconstructed in an event, opposite-sign track pairs are combined if they satisfy basic track quality criteria. Topological requirements are applied to separate the conversions from random opposite-sign track pairs. These requirements include cuts on the transverse impact parameter, on the distance of minimum approach in the transverse plane, and on the opening angle in the longitudinal plane [28]. The surviving track pairs are then used in a 3-D constrained kinematic vertex fitter, where the invariant mass is constrained to be 0, and the tracks are imposed to be parallel at the CV. Convergent fits with reasonable $\chi^{2}$-probabilities finally lead to photon conversion candidate objects that are saved in the event, and that can be further used in the physics analyses. These conversion objects will be referred to as "general conversions". Figure 3.19 shows the distribution of conversion vertices obtained from

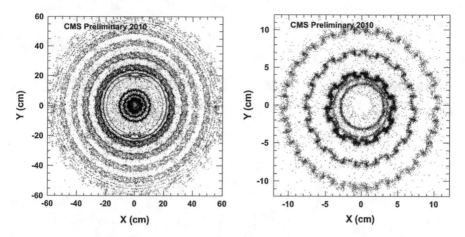

**Fig. 3.19** Distribution of the conversion vertices in the transverse plane, for $|z| < 26$cm, with different levels of zoom in $x$ and $y$, increasing from *left* to *right* [28]

general conversions in the transverse plane, clearly identifying the individual tracker layers [28].

Thanks to the excellent performance of the silicon tracker, the mass resolution of the $\chi$ candidates built with photons reconstructed using this technique is of the order of 6 MeV, for the $\chi_c$ and $\chi_b(1P)$ systems. Moreover, the selection requirements ensure that the $\chi$ mass spectrum only suffers from a small background contribution. The disadvantage of this approach is on one hand the low conversion probability, resulting by itself in a loss of the order of 70% of the events, as shown by simulation, and on the other hand the low photon conversion reconstruction efficiency of the algorithms and requirements explained above. The conversion reconstruction efficiency, convoluted with the conversion probability, is shown in Fig. 3.20, as a function of the photon $p_T$. This quantity reaches a plateau at around 5 GeV, and decreases steeply at lower $p_T$.

The biggest loss of efficiency is caused already at the seeding step. Obviously, if one or both of the electron tracks are not seeded, the corresponding conversion cannot be reconstructed. For the first case, when one of the electron tracks is reconstructed but the second electron track was not seeded, an efficient solution was found and implemented in the general conversion finding sequence, the so-called "single-leg" seeding. After the iterative tracking, individually for each reconstructed track, a search for a hit-pair is started, building the trajectory with the assumption that the hit-pair was produced by a conversion electron whose trajectory has a tangent point with the already reconstructed track. This algorithm evidently improves the conversion reconstruction efficiency, especially at high conversion radii and for low-$p_T$ tracks. Due to the nature of the algorithm, starting from an already reconstructed track, the processing time is rather short. However, if both electron tracks are missed by the default tracking iterations, the single-leg approach does not improve the conversion reconstruction. A possible solution to this problem is discussed in detail below.

**Fig. 3.20** Convolution of conversion probability and conversion reconstruction efficiency, as a function of the photon $p_T$, estimated from simulation for CMS data taking conditions as in the 2011 run [29]

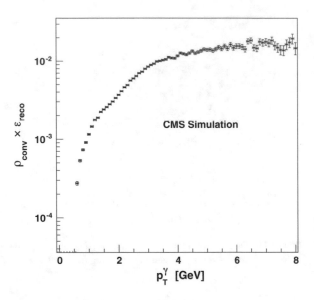

## Improvement of Low-Energy Photon Conversion Reconstruction

A different seeding approach has been studied as an improved solution for when both electron tracks are missed by the general CMS reconstruction. Instead of attempting to seed two independent tracks, which in the end can be identified by the conversion finder as a general conversion, the idea is to identify a clear hit pattern corresponding to a conversion, and seeding directly and simultaneously the tracks of the two electrons, constrained by the physics information about the process. This new seeding approach, denoted here as "quad-seeding", was originally proposed by Giacomo Sguazzoni and Domenico Giordano, then studied by Evan Song, who also implemented the geometrical solution of the problem. The final implementation in the CMSSW environment was done by myself after optimizing the code, the selection cuts, and after conducting performance studies with simulation data, in collaboration with Wolfgang Adam.

Given the geometry of the magnetic field, the problem can be simplified to a 2-dimensional problem by projecting the conversion process onto the transverse plane, described through the polar coordinates $r$ and $\varphi$, the nominal interaction point being the pole of the coordinate system. The BS is assumed to correspond to the point of origin of the photon in the general case, where no information about the PV is available. The photon converts at the CV, defined by $r_{CV}$, the conversion radius and $\varphi_{CV}$. The electrons, parallel at the CV, are bent by the Lorentz force in a circular path, as seen in the transverse plane, with radii $R_1$ and $R_2$, and leave hits in the silicon detector layers they cross.

In order to unambiguously determine the position of the CV in the transverse plane, as well as the radii of the two electron trajectories, four hits are sufficient, given the constraint from the PV. Figure 3.21 shows the geometry of the problem

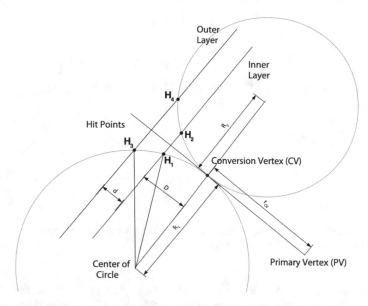

**Fig. 3.21** Geometry of a photon conversion, as seen in the transverse plane, seeded by a quad-seed – two hits on two detector layers each [30]. Refer to the discussion in the text for further details

as discussed here. Combinations with two hits from each electron on two layers each are considered: $H_1$, $H_3$ and $H_2$, $H_4$ on the "inner" and "outer" detector layers, with radial coordinates $r_{H_{1,2}}$ and $r_{H_{3,4}}$, respectively. The solution of this geometrical problem leads to coupled quadratic equations that are difficult to solve analytically. A robust and efficient iterative numerical solution was developed, which was optimized and validated with simulated pseudo-data samples [30].

**Quad-seeds: Optimization and Performance**

The main problem of this new seeding approach is the huge number of seeds when considering all four-hit combinations with two hits on two layers each. In some cases, as found in this study, more than $10^7$ combinations per event have to be considered, certainly too much for a practical algorithm, for which optimization of time consumption is required. In order to save CPU time, seed cleaning cuts are defined that can be applied to these hit combinations before building the seeds, as for example:

- Cuts on the "Intersection point" (ISP): As sketched in Fig. 3.22 (left), the ISP is defined in the transverse plane as the intersection of the two lines containing $H_1$, $H_3$ and $H_2$, $H_4$, respectively. As at this stage, no information about the CV is available. However, the radial coordinate of the ISP, $r_{ISP}$, constitutes an upper bound for $r_{CV}$. In order to avoid tracks originating from within the beam pipe, combinations with $r_{ISP} < 3$ cm are rejected. Furthermore, the best results are expected for hits from detector layers close to the CV. Therefore, hit combinations for which the difference between $r_{H_{1,2}}$ and $r_{ISP}$ is larger than 20 cm are rejected.

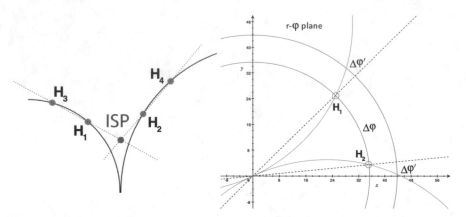

**Fig. 3.22** Sketches explaining the geometry of the seed cleaning cuts on the intersection point [31] (*left*) and on $\Delta\varphi$ (*right*), in the transverse plane

- Topological cuts on $\Delta\varphi$, in the transverse plane: The geometry of these cuts is sketched in Fig. 3.22 (right). Assuming the CV to coincide with the interaction point and requiring a minimum $p_T$ for both electron legs of 100 MeV (represented by the curvature of the green lines), starting from $H_1$ one can define a $\Delta\varphi$ region within which $H_2$ is searched for. Moreover, extrapolating the trajectory to the outer layer, with the same criterion one obtains the angular region $\Delta\varphi'$, within which the hits $H_3$ and $H_4$ are searched for. Hit combinations that do not fulfill the $\Delta\varphi$ and $\Delta\varphi'$ criteria are rejected [30, 31].

With these cuts, the number of quad-seed candidates per event can be reduced considerably, without affecting significantly the overall performance. After the quad-seeding algorithm has identified the individual quad-seeds of a given event, including information about the position of the CV, the seeds for which the difference between $r_{H_{1,2}}$ and $r_{CV}$ is larger than 20 cm can be rejected. Furthermore, a simple arbitration algorithm can be applied, in order to avoid propagating seeds to the tracking algorithms that share a subset of the hits.

Detailed studies are available concerning the impact of the individual cuts. In the study presented here, non of these cuts are applied. As a first step, it is important to understand what is the maximum possible gain in conversion reconstruction efficiency when including the quad-seeding approach in addition to the standard CMS conversion reconstruction. This study is restricted to the central region, $|\eta|^\gamma < 1.2$, as in this region the best results are expected. This study is performed using simulated data resembling 2012 data taking conditions. Two different kinds of MC samples have been used. Firstly, a simple "particle-gun" MC has been produced, each event containing one photon, generated flat in $p_T^\gamma$, covering the range $0.15 < p_T^\gamma < 10$ GeV. This sample is referred to as "Single Photon" (SP) sample, containing roughy 10 k reconstructed general conversions. The second sample is more complex and realistic, generated with PYTHIA [32], simulating a realistic event content of $pp$ collisions at $\sqrt{s} = 8$ TeV, though, not including the simulation of PU collisions. This sample is referred to as "Minimum Bias" (MB) sample, containing roughly 30 k reconstructed general conversions in the region of interest.

With these samples, detailed performance studies can be conducted, comparing the reconstruction performance with and without employing the quad-seeding approach. The interpretation of these studies require the definition and evaluation of the efficiency and fakerate, on "seed level" and on "conversion level". On seed level, all identified quad-seeds are matched to the truth information of a given MC event. Similarly, on conversion level, the seeds that were used to successfully seed a given reconstructed conversion are matched to the truth information of the MC event. In both cases, the seeds/conversions are only matched, if both electron legs are successfully associated. Efficiency and fakerate are defined as

$$\epsilon_{seed} = \frac{\text{matched quad-seeds}}{\text{MC truth conversions}} , \quad f_{seed} = 1 - \frac{\text{matched quad-seeds}}{\text{quad-seeds}} ,$$

$$\epsilon_{conv} = \frac{\text{matched reco-conversions}}{\text{MC truth conversions}} , \quad f_{conv} = 1 - \frac{\text{matched reco-conversions}}{\text{reco-conversions}} .$$

Before entering the denominator of the efficiency calculations, both legs of the MC truth conversions are required to lead to at least 4 hits in the tracker, to ensure that both electron legs can be reconstructed.

Figure 3.23 summarizes the performance studies. The top panel shows the distribution of the number of identified quad-seeds per event, for the SP and the MB samples. This clearly shows (as expected) that in MB events the number of identified quad-seeds is much larger than in SP events. The diagrams in the second row show the results for $\epsilon_{seed}$ as a function of the electron $p_T^e$ (left) and as a function of the electron $\eta^e$ (right). As expected, an increasing trend with $p_T^e$ is observed. In the SP case, the seed efficiency reaches a value of around 60% at $p_T^e = 1$ GeV while in the MB case the seed efficiency is around 25%. The seed efficiency is best at mid-rapidity, and decreases with $\eta^e$. The diagrams in the third row show the results for $\epsilon_{conv}$ as a function of $p_T^e$ (left) and $\eta^e$ (right), comparing the conversion efficiency of the MB and SP studies, both with and without employing the quad-seed algorithm. At $p_T^e = 1$ GeV, the conversion efficiency of the SP sample analysis is around 13.5 and 23%, without and with using the quad-seed approach. The corresponding values of the MB analysis are around 10 and 12%. Studies have shown that the conversion fakerate $f_{conv}$ does not significantly change after adding the quad-seed approach. In order to better appreciate the impact of using the quad-seed feature, the bottom panels show the ratio of the conversion efficiency after using the quad-seeds with respect to the standard reconstruction algorithms. At very low values of $p_T^e$, below 500 MeV, the conversion efficiency increases by around 240% for the SP analysis and by around 50% for the MB analysis. The efficiency-gain is found to be especially large for conversions with high conversion vertex radii, $r_{CV} > 20$ cm. The reconstruction of photons that convert in the TIB and TOB is very inefficient with the standard CMS conversion reconstruction software, and can be significantly improved with the quad-seed approach. The large difference in performance between the SP and MB studies shows that the number of tracks (and therefore quad-seeds) in

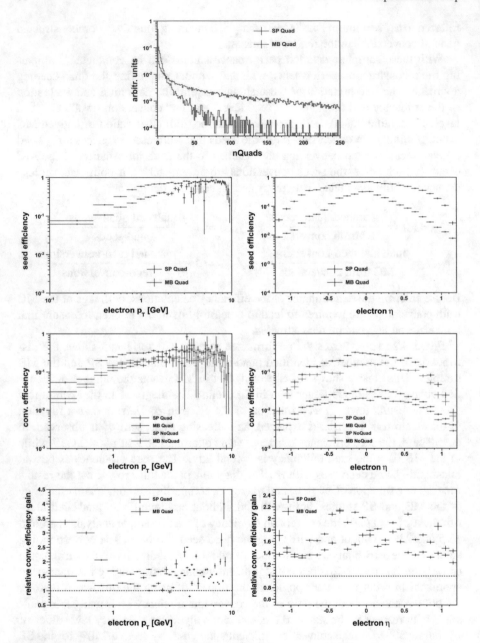

**Fig. 3.23** Summary of the quad-seed performance studies: distribution of the number of identified quad-seeds per event (*top*); seed efficiency (*second row*), conversion efficiency (*third row*) and conversion efficiency gain (*bottom row*), as a function of $p_T^e$ (*left*) and $\eta^e$ (*right*)

a given event has a large influence on the performance of the quad-seed conversion reconstruction. Contrary to the MB sample, data are affected by PU, resulting in an increased number of tracks and quad-seeds, probably reducing the efficiency-gain that can be obtained when using the quad-seeding approach.

In conclusion, the quad-seeding approach leads to a considerable gain in conversion efficiency at low $p_T^e$ and high $r_{CV}$, which makes it an interesting feature for P-wave quarkonium analyses, given the soft photon momentum distribution of their radiative decays. However, the overall performance is not as good as expected, so that the usage of the time-consuming quad-seed approach within the general CMS reconstruction software is not feasible. The code is therefore implemented in the official CMSSW releases, but deactivated in the default reconstruction. Even though quad-seeds will not enter the CMS standard reconstruction, there are possible specific applications of the quad-seed feature in the future, as for example in P-wave quarkonium analyses. The presence of information about the dimuon vertex, which is identical with the photon vertex, further improves the quad-seed performance, given that without PV information, the photon origin has to be assumed to coincide with the BS, affected by very large uncertainties. Therefore, one can "re-reconstruct" the data including quad-seeds, in view of increasing the statistics of quarkonium P-wave analyses. This re-reconstruction is possible with existing data collected in Run I, as well as with future data, to be taken in Run II.

### 3.2.7 Quarkonium Reconstruction Performance

This section briefly discusses the performance of the CMS experiment regarding the quarkonium physics program.

**S-Wave Quarkonia**

The S-wave quarkonium states are reconstructed in the very clean dimuon decay channel, driven by very efficient dimuon physics triggers, only limited by the allocated bandwidth and the acceptance for low-$p_T$ muons. One important aspect of the physics impact of a measurement of quarkonium production cross sections and polarizations is the high-$p_T$ reach. Given that CMS can profit from the full LHC luminosity and has an efficient and flexible trigger system, as well as very efficient muon reconstruction, it has the possibility to access the highest-possible values of $p_T$, only limited by the conditions the LHC provides. Moreover, the geometrical coverage allows CMS quarkonium physics analyses to access the mid-rapidity range. LHCb, for example, was limited in Run I to quarkonium cross section measurements up to around 15 GeV [33–36], while CMS was able to perform the same measurements up to around 100 GeV [37, 38].

Another important aspect of quarkonium production measurements is the possibility to handle the mass continuum background. Besides the excellent muon identification, CMS has a very good resolution of the dimuon mass, $M_{\mu\mu}$. The fraction of background events under the signal peak can be kept reasonably small, given that this

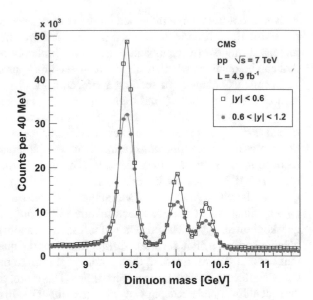

**Fig. 3.24** Dimuon mass
distribution of events with
$p_T > 10\,\text{GeV}$ in the $\Upsilon(nS)$
mass region, in two ranges of
$|y|$ [40]

fraction is proportional to the mass resolution. Moreover, many analyses rely on modeling the background under the mass signal peaks by interpolation of the distributions of the "mass sidebands", the regions above and below the signal peaks, containing negligible signal contributions. The systematic uncertainties associated to this interpolation decrease with improving mass resolution, given that the sidebands can be defined to be closer to the peak region. The CMS dimuon mass resolution is slightly worse than in LHCb [36], but slightly better than in ATLAS [39]. CMS can measure the dimuon mass with a relative resolution of around 0.6–1.4%, depending on the dimuon kinematics. The mass resolution is best for low-$p_T$ and mid-rapidity, worst for high-$p_T$ forward-rapidity events. Contrary to the ATLAS measurements [39], the CMS mass resolution enables to very well separate the $\Upsilon(2S)$ and $\Upsilon(3S)$ states, except for the high-$p_T$ forward-rapidity region, where the two states are affected by some overlap. Figure 3.24 illustrates the dimuon mass reconstruction capabilities of CMS through an example distribution in the $\Upsilon(nS)$ mass region.

One more important aspect, relevant for charmonium production analyses, is the capability to take into account the feed-down decays of B hadrons, denoted as the "non-prompt" (NP) contribution, which can constitute the majority of the charmonium data samples, especially at high $p_T$ [41, 42]. The individual B hadrons have average decay times of the order of $10^{-12}$ s [26], allowing them to travel a certain average distance of the order of $10^2$ μm through the detector before decaying into $\psi(nS) + X$, which can be resolved by the excellent CMS silicon tracker. Therefore, it is possible to discriminate between prompt and non-prompt charmonia on a

**Fig. 3.25** Simplified sketch
of the definition of the most
probable transverse decay
length $L_{xy}$, illustrating a
decay chain
$B \rightarrow \psi(nS) + X$,
$\psi(nS) \rightarrow \mu\mu$ in the $r$-$z$ plane

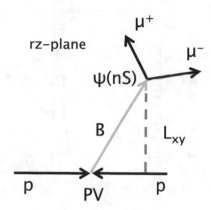

statistical basis,[4] using the so-called "pseudo-proper lifetime"[5] [43] variable (here
simply referred to as the "lifetime"),

$$\ell = L_{xy} \cdot M_{\mu\mu}/p_T^{\mu\mu} \; , \tag{3.5}$$

with $L_{xy}$ the most probable transverse decay length in the laboratory frame,

$$L_{xy} = \frac{\mathbf{u}^T \sigma^{-1} \mathbf{x}}{\mathbf{u}^T \sigma^{-1} \mathbf{u}} \; , \tag{3.6}$$

where $\mathbf{x}$ is the vector in the transverse plane connecting the dimuon vertex and the
primary vertex of the event, $\mathbf{u}$ is the unit vector of the $\psi(nS)$ $p_T$, and $\sigma$ is the sum of
the covariance matrices of the two vertices under consideration. Figure 3.25 shows
a sketch of the definition of $L_{xy}$. The CMS tracker can resolve the primary and
dimuon vertices with a precision that results in a resolution of the lifetime variable,
$\sigma_\ell$, between 10 and 30 μm, best at high $p_T$ and worst at low $p_T$. This allows us
to study the lifetime distributions of the prompt and non-prompt components in
a straightforward way and, for example, to define regions in lifetime where the
contamination of non-prompt events can be reduced by a factor of up to 4, without
rejecting a significant amount of prompt signal events (see Sect. 4.3). Figure 4.17 in
Sect. 4.3.3 nicely illustrates the lifetime reconstruction capabilities of CMS through
example distributions in the $\psi(nS)$ mass regions.

It has to be emphasized that, while the non-prompt feed-down decays can be
very well separated from the prompt component in CMS analyses, it is not possible
to separate the directly produced quarkonia from the feed-down decays of heavier quarkonium states. This is true for all LHC experiments. The measurements
of quarkonium production are limited to the measurement of the properties of the

---

[4]Given that the modes of the decay time distributions of the prompt and non-prompt charmonia are
both at 0, they can not be distinguished on an event-by-event basis.

[5]The word "pseudo" refers to the fact that the kinematics of the decaying B hadrons are not reconstructed, but approximated by the dimuon kinematics.

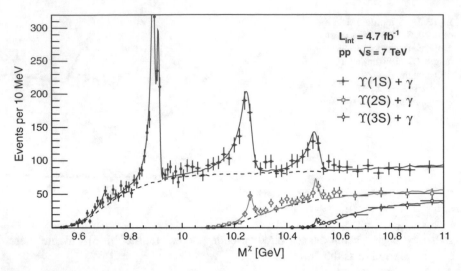

**Fig. 3.26** Mass spectrum of the $\Upsilon(nS) + \gamma$ decay channels, as measured by CMS at 7 TeV, including a fit to the three distributions

prompt component. However, through measurements of the same observables for the heavier quarkonium states, and the respective feed-down fractions, the corresponding observables of the directly produced quarkonia are accessible a posteriori.

**P-Wave Quarkonia**

Concerning the performance of the CMS detector with respect to the analyses of P-wave production and polarization, many considerations stated above are also valid in this case, given that the dimuon reconstruction is the first step of the P-wave reconstruction, with the additional requirement of a reconstructed photon. The capabilities to take into account the non-prompt B-hadron decays are the same for P-wave analyses as for S-wave analyses, given that the dimuon vertex displacement can be calculated in the same way, independently of the presence of the photon.

The $p_T$ reach of P-wave measurements is statistically limited, given that the photon conversion reconstruction efficiency is rather low. Nonetheless, the $p_T$ reach of CMS in these measurements is still beyond all other experiments, only ATLAS presenting comparable performances. While the CMS photon conversion reconstruction efficiency is small, the corresponding fakerate is small as well.

The CMS capabilities regarding P-wave analyses are well illustrated by Fig. 3.26, which shows the $\Upsilon(nS) + \gamma$ ($n = 1, 2, 3$) mass spectra, $M^\chi$, as measured by CMS at 7 TeV, as well as the result of a simultaneous unbinned ML fit to the three distributions. The fitting strategy is briefly discussed here. The shape of the background continuum under the signal peaks is built using a data-driven procedure, mixing the kinematical properties of the dimuon of a given event with the kinematical properties of the photon of a different event, randomly selected. These two objects are combined to form a sample of pseudo $\chi_b$ candidates, representing the uncorrelated combinatorial background contribution [44]. The $\chi_b(1P)$, $\chi_b(2P)$ and $\chi_b(3P)$ peaks are modeled as a doublet of Crystal-Ball (CB) functions [45] (see Sect. 4.2.3 for more details), representing a superposition of the J=1 and J=2 states, neglecting the J=0 states. The relative weights $\chi_{b2}(nP)/\chi_{b1}(nP)$ for the three doublets, as well as the mass differences of the $\chi_{b1}(nP)$ and $\chi_{b2}(nP)$ states, are fixed to reasonable values, inspired by Refs. [26, 46, 47]. Besides the peaks of the $\chi_b(nP) \rightarrow \Upsilon(1S) + \gamma$ (n=1, 2, 3), $\chi_b(nP) \rightarrow \Upsilon(2S) + \gamma$ (n=2, 3) and $\chi_b(3P) \rightarrow \Upsilon(3S) + \gamma$ decays, a so-called "reflection" peak is taken into account at around $M^\chi = 9.7\,\text{GeV}$, which originates from the combination of the photon from the decay $\chi_b(2P) \rightarrow \Upsilon(2S) + \gamma$ with the dimuon from the $\Upsilon(1S)$ following the decay $\Upsilon(2S) \rightarrow \Upsilon(1S)\pi\pi$.

The resolution of $M^\chi$ is excellent, because the dimuon mass resolution cancels due to the way $M^\chi$ is defined: $M^\chi = M_{\mu\mu\gamma} - M_{\mu\mu} + M^{PDG}_{\Upsilon(nS)}$. The $M^\chi$ mass resolution depends on the photon energy, and therefore on the mass difference of the corresponding $\chi$ and S-wave states involved in the decay, and is of the order of 6 MeV for the $\chi_c \rightarrow J/\psi + \gamma$, $\chi_b(1P) \rightarrow \Upsilon(1S) + \gamma$, $\chi_b(2P) \rightarrow \Upsilon(2S) + \gamma$ and $\chi_b(3P) \rightarrow \Upsilon(3S) + \gamma$ decays, characterized by rather small mass differences between the mother- and daughter-quarkonia. The average photon energies are proportional to the mass differences and therefore, for the $\chi_b(2P) \rightarrow \Upsilon(1S) + \gamma$, $\chi_b(3P) \rightarrow \Upsilon(2S) + \gamma$ and $\chi_b(3P) \rightarrow \Upsilon(1S) + \gamma$ decays, the $M^\chi$ mass resolutions are worse. The mass resolution allows the separation of the J=1 and J=2 states in the $\chi_c$, $\chi_b(1P)$ and $\chi_b(2P)$ systems, maybe even for the $\chi_b(3P)$ system, depending on the not yet measured mass splitting of the spin-triplet states. The CMS capabilities of the separation of the J=1 and J=2 states in the $\chi_b(1P)$ system can be seen in Fig. 3.26, already exploited in the CMS measurement of the relative $\chi_{b2}(1P)/\chi_{b1}(1P)$ production cross section ratios at 8 TeV [48]. The J=0 states are more difficult to access experimentally, given the branching fractions of the respective radiative decays, which are very small, in the cases where they are measured [26]. Moreover, the natural width of the $\chi_{c0}$ is larger than the $M^\chi$ mass resolution, such that it will always be a broad peak, accumulating more background underneath.

For the selection of P-wave quarkonium candidates, one can exploit constraints on both $M_{\mu\mu}$ as well as the invariant mass of the $\mu\mu\gamma$ system, $M^\chi$, in order to reduce the background contamination. The level of background under the $\chi$ signal peaks is generally rather small, especially when compared to the other LHC experiments. The high-$p_T$ reach and excellent $M^\chi$ resolution provide CMS with the opportunity to perform measurements in the quarkonium P-wave sector that are highly competitive with ATLAS and LHCb.

## 3.3 Experimental Setup Summary

The performance of the LHC accelerator was above expectations throughout Run I, providing excellent conditions for data taking at the CMS experiment. Thanks to its high collision energies and integrated luminosities, the LHC can be regarded as a "quarkonium factory", producing S-wave and P-wave quarkonium states with rates never seen in previous hadron colliders.

The CMS experiment has also performed very well in the Run I data taking period. The trigger system works very well, with a higher output event rate than anticipated in the nominal design, improving the physics output of the experiment. The silicon tracker and muon systems function very well, providing accurate and efficient reconstruction of high level physics objects for data analyses.

Quarkonium cross sections and polarizations can be studied with the CMS detector with an excellent performance, given the excellent muon identification, the efficient trigger system, and the very good dimuon mass, dimuon lifetime and $\mu\mu\gamma$ mass resolutions. In conclusion, CMS has better capabilities than the other LHC experiments in many aspects relevant for quarkonium production measurements, for both S-wave and P-wave quarkonium states.

# References

1.  Evans L, Bryant P (2008) LHC machine. J Instrum 3:S08001
2.  CMS Collaboration (2006) CMS Physics Technical Design Report Volume I: Detector Performance and Software. CERN-LHCC-2006-001
3.  CMS Collaboration (2007) CMS Physics Technical Design Report, Volume II: Physics Performance. J Phys G Nucl Part Phys 34:995
4.  CMS Collaboration (2008) The CMS experiment at the LHC. J Instrum 3:S08004
5.  Lefèvre, C (2008) The CERN accelerator complex. Complexe des accélérateurs du CERN. https://cdsweb.cern.ch/record/1260465?ln=en
6.  CERN (2014) LHC-Statistics. http://lhc-statistics.web.cern.ch/LHC-Statistics/
7.  CMS Collaboration (2012) Observation of a new boson at a mass of 125 GeV with the CMS experiment at the LHC. Phys Lett B 716:30
8.  ATLAS Collaboration (2012) Observation of a new particle in the search for the Standard Model Higgs boson with the ATLAS detector at the LHC. Phys Lett B 716:1
9.  CMS Collaboration (2013) CMS Public Luminosity Results. https://twiki.cern.ch/twiki/bin/view/CMSPublic/LumiPublicResults
10. Martini L (2014) Measurement of the $B_0^s \to \mu^+\mu^-$ branching fraction and search for the $B_0 \to \mu^+\mu^-$ decay with the CMS experiment. Ph.D. thesis, Universita degli Studi di Siena
11. Barney D (2012) CMS slice image with transverse view. https://cms-docdb.cern.ch/cgi-bin/PublicDocDB/ShowDocument?docid=5697
12. CMS Collaboration (2014) Description and performance of track and primary-vertex reconstruction with the CMS tracker. CMS-TRK-11-001
13. CMS Collaboration (2000) CMS. The TriDAS project. Technical design report, vol. 1: The trigger systems. CERN-LHCC-2000-038
14. CMS Collaboration (2010) CMS Muon Results. https://twiki.cern.ch/-twiki/bin/view/-CMS-Public/PhysicsResultsMUO
15. CMS Collaboration (2011) CMS B-Physics and Quarkonia Results. https://twiki.cern.ch/twiki/bin/view/CMSPublic/PhysicsResultsBPH
16. CMS Collaboration (2014) CMS web based monitoring. https://cmswbm.web.cern.ch/cmswbm/
17. Brun R, Rademakers F (1997) ROOT: an object oriented data analysis framework. Nucl Instrum Meth A389:81
18. CMS Collaboration (2010) CMS tracking performance results from early LHC operation. Eur Phys J C 70:1165

19. Private Communication from B. Mangano
20. Cucciarelli S, Konecki M, Kotlinski D, Todorov T (2006) Track reconstruction, primary vertex finding and seed generation with the Pixel Detector. CMS-NOTE-2006-026
21. Frühwirth R (1987) Application of Kalman filtering to track and vertex fitting. Nucl Instrum Meth A 262:444
22. Adam W, Mangano B, Speer T, Todorov T (2006) Track Reconstruction in the CMS tracker. CMS-NOTE-2006-041
23. CMS Collaboration (2010) Tracking and Vertexing Results from First Collisions. CMS-PAS-TRK-10-001
24. CMS Collaboration (2009) Algorithms for b Jet Identification in CMS. CMS-PAS-BTV-09-001
25. CMS Collaboration (2012) Performance of CMS muon reconstruction in $pp$ collision events at $\sqrt{s} = 7$ TeV. CMS-PAS-MUO-10-004
26. Olive KA et al (2014) Particle data group. Chin Phys C 38:090001
27. Klein SR (2006) e+ e- pair production from 10-GeV to 10-ZeV. Radiat Phys Chem 75:696
28. CMS Collaboration (2010) Studies of Tracker Material. CMS-PAS-TRK-10-003
29. CMS Collaboration (2012) CMS BPH-11-010 Supplemental Material. https://twiki.cern.ch/twiki/bin/view/CMSPublic/PhysicsResultsBPH11010
30. Private Communication from E. Song
31. Private Communication from G. Sguazzoni
32. Sjostrand T, Mrenna S, Skands PZ (2006) PYTHIA 6.4 physics and manual. J High Energy Phys 0605:026
33. LHCb Collaboration (2011) Measurement of $J/\psi$ production in $pp$ collisions at $\sqrt{s} = 7$ TeV. Eur Phys J C 71:1645
34. LHCb Collaboration (2012) Measurement of $\psi(2S)$ meson production in pp collisions at $\sqrt{s} = 7$ TeV. Eur Phys J C 72:2100
35. LHCb Collaboration (2012) Measurement of Upsilon production in pp collisions at $\sqrt{s} = 7$ TeV. Eur Phys J C 72:2025
36. LHCb Collaboration (2013) Production of $J/\psi$ and $\Upsilon$ mesons in $pp$ collisions at $\sqrt{s} = 8$ TeV. J High Energy Phys 1306:064
37. CMS Collaboration (2015) Measurement of prompt $J/\psi$ and $\psi(2S)$ double-differential cross sections in $pp$ collisions at $\sqrt{s} = 7$ TeV. Submitted to Phys Re Lett. arXiv:1502.04155
38. CMS Collaboration (2015) Measurements of the $\Upsilon(1S)$, $\Upsilon(2S)$ and $\Upsilon(3S)$ differential cross sections in $pp$ collisions at $\sqrt{s} = 7$ TeV. Submitted to Phys Lett B. arXiv:1501.07750
39. ATLAS Collaboration (2013) Measurement of Upsilon production in 7 TeV pp collisions at ATLAS. Phys Rev D 87:052004
40. CMS Collaboration (2013) Measurement of the $\Upsilon(1S)$, $\Upsilon(2S)$ and $\Upsilon(3S)$ polarizations in $pp$ collisions at $\sqrt{s} = 7$ TeV. Phys Rev Lett 110:081802
41. CMS Collaboration (2012) $J/\psi$ and $\psi(2S)$ production in $pp$ collisions at $\sqrt{s} = 7$ TeV. J High Energy Phys 1202:011
42. ATLAS Collaboration (2011) Measurement of the differential cross-sections of inclusive, prompt and non-prompt $J/\psi$ production in proton-proton collisions at $\sqrt{s} = 7$ TeV. Nucl Phys B 850:387
43. CMS Collaboration (2011) Prompt and non-prompt $J/\psi$ production in pp collisions at $\sqrt{s} = 7$ TeV. Eur Phys J C 71:1575
44. Private Communication from E. Aguiló
45. Oreglia MJ (1980) A Study of the Reactions $\psi(2S) \rightarrow \gamma\gamma\psi$. Ph.D. thesis, SLAC-R-236:Appendix D
46. CMS Collaboration (2012) Measurement of the relative prompt production rate of $\chi_{c2}$ and $\chi_{c1}$ in $pp$ collisions at $\sqrt{s} = 7$ TeV. Eur Phys J C 72:2251
47. Motyka L, Zalewski K (1998) Mass spectra and leptonic decay widths of heavy quarkonia. Eur Phys J C 4:107
48. CMS Collaboration (2014) Measurement of the production cross section ratio $\sigma(\chi_{b2}(1P))/\sigma(\chi_{b1}(1P))$ in $pp$ collisions at $\sqrt{s} = 8$ TeV. Submitted to Phys Lett B. arXiv:1409.5761

# Chapter 4
# Data Analysis

This chapter describes in detail the measurements of quarkonium polarization with the CMS detector. The general analysis strategy, applicable to all the CMS quarkonium polarization analyses, is described in Sect. 4.1, while the details of the individual analyses are described in Sects. 4.2 and 4.3, for the analysis of the polarizations of the $\Upsilon(nS)$ and prompt $\psi(nS)$ states, respectively. These measurements were conducted in a team effort, involving undergraduate and graduate students, as well as senior scientists, from several institutions, including HEPHY (Vienna, Austria), CERN (Geneva, Switzerland), LIP (Lisbon, Portugal) and PKU (Beijing, China).

The analyses described in this chapter were published in Refs. [1, 2]. The analysis strategies, the required inputs, discussions of relevant systematic uncertainties and results are documented in detail in internal CMS notes [3–7], and were scrutinized extensively by the full CMS Collaboration. This chapter summarizes the most important aspects of these measurements, with a focus on the parts of the analyses that are characterized by significant contributions of myself.

## 4.1 Analysis Strategy

The CMS measurements of quarkonium polarization heavily rely on improved analysis techniques (with respect to the pre-LHC era), suggested by a series of papers [8–13], as discussed in Sect. 2.3. Therefore, the full angular distributions are measured, including the anisotropy parameters $\lambda_\vartheta$, $\lambda_\varphi$ and $\lambda_{\vartheta\varphi}$, in three reference frames, the Collins-Soper, Helicity and Perpendicular Helicity frames. In addition, we also measure the frame-invariant parameter $\tilde{\lambda}$, which constitutes both an interesting physics observable for the interpretation of the results, as well as an important experimental cross-check of the overall analysis strategy, by comparing the obtained results of the frame-invariant parameter $\tilde{\lambda}$ as measured in different reference frames.

© Springer International Publishing Switzerland 2017
V. Knünz, *Measurement of Quarkonium Polarization to Probe QCD
at the LHC*, Springer Theses, DOI 10.1007/978-3-319-49935-2_4

Given the possibility that the anisotropy parameters depend on the quarkonium production kinematics (see Chap. 2), the polarization parameters $\vec{\lambda}$ are measured in small bins of the transverse momentum $p_T$ of the quarkonium states, as well as in several ranges in rapidity $|y|$. This has the additional advantage of not integrating events with substantially different kinematics, which could potentially introduce biases in the analyses of the decay angular distributions [12].

The analysis of quarkonium polarizations is a challenging task, and has to be conducted with great care, in order to avoid a continuation of the series of experimental inconsistencies in the pre-LHC era (see Sect. 2.2.2). The complexity of quarkonium polarization analyses can in part be explained by the multi-dimensionality of the problem. Besides the angular dimensions $\cos \vartheta$ and $\varphi$, the dimuon invariant mass $M_{\mu\mu}$ has to be studied, in order to take into account the continuum $\mu\mu$ background. In the case of the measurements of charmonium polarizations, additionally the pseudo-proper lifetime, $\ell$, has to be taken into account, in order to subtract the non-prompt charmonium contributions, originating from decays of heavier B hadrons, distinguishable due to their average lifetimes of the order of $10^{-12}$ s, resolvable with the precision of the CMS silicon tracker. In the following, all contributions besides the prompt quarkonium signal are denoted as "inclusive background", including the $\mu\mu$ background and the non-prompt charmonium signal (in the case of charmonium analyses).

The most critical aspects of the measurements are briefly introduced here:

- One of the most challenging aspects of quarkonium polarization measurements is the severe restriction in angular acceptance, imposed by cuts on low-$p_T$ muons and by the low-$p_T$ muon efficiencies. Both effects lead to a depopulation of events close to $|\cos \vartheta^{PX}| = 1$, the region in angular phase space that has the largest power to distinguish between transverse and longitudinal distributions, therefore having a decisive influence on the precision of the results. These effects are most prominent in low dimuon $p_T$ regions, limiting the measurements to regions above a certain $p_T^{min}$, which is different for the individual analyses.
- Another challenge is the precise mapping of the muon efficiencies. The so-called "turn-on curve", a region at low muon $p_T$ where the efficiencies increase very quickly as a function of $p_T$, has to be known with great precision, as biases in the efficiency determination can introduce artificial polarizations, biasing the measurement. Furthermore, the cases where the two decay muons pass through the CMS muon stations close-by are affected by severe inefficiencies, mostly caused by the trigger algorithms. This is the case for high-$p_T$, low-mass quarkonia, affecting the charmonium analyses in a more serious way than the bottomonium analyses. These muon-pair correlations lead to a depopulation of events close to $|\cos \vartheta^{PX}| = 0$, introducing an artificial transverse polarization in the PX and HX frames, that has to be corrected for.
- The systematic uncertainties regarding the angular and kinematic distributions of the background have to be evaluated with great care. The background distributions are modeled from the data in the signal-depleted mass sidebands, and are interpolated under the signal peaks.

The framework used for the extraction of the anisotropy parameters was developed, optimized and validated in collaboration with Pietro Faccioli and João Seixas, documented in detail in a CMS internal note [3], and summarized in the remaining part of this section.

## 4.1.1 The Polarization Extraction Framework

The traditional approaches to the study of quarkonium polarization, employing ML fits using minimization tools such as MINUIT [14], lead to several problems of instability, as shown by toy experiments, discussed in the CMS internal note [15], mainly due to the restrictions of the phase space in the angular variables. This finding has led to the development of new techniques, both in the modeling of the physics problem, as well as in numerical techniques that are employed, resulting in a method of extracting the anisotropy parameters from data in a stable and reliable way.

The framework described in this section is applicable to the measurement of the polarization of any quarkonium state. It does not rely on any MC acceptance maps or any other template methods. Therefore, no assumptions on any theoretical model enter the method of extracting the polarization parameters. The only external inputs required by the framework are the dimuon efficiencies, which are obtained by the data-driven "tag and probe" ($T\&P$) method [16]. However, the CMS detector suffers from muon-pair correlations affecting the dimuon efficiencies, whose effect on the angular distributions has to be studied with MC, validated by studies on data. The determination of the muon and dimuon efficiencies is described in more detail in Sects. 4.2.2 and 4.3.2.

The data-driven philosophy of the CMS analyses is also followed in the background subtraction procedure, which is a novel technique, subtracting "background-like" events from the data sample, on an event-by-event basis, based on a likelihood ratio criterion. The information that is used in the background subtraction algorithm includes the full dilepton kinematics ($p_T$, $|y|$, $M_{\mu\mu}$, $\cos\vartheta$, $\varphi$) of the background events, estimated from the dimuon mass sidebands of the data sample itself. The background subtraction algorithm is described in more detail in Sect. 4.1.2.

It should be emphasized that (as explained in more detail in Sect. 4.1.3) the method developed for the CMS quarkonium polarization analyses is not a fitting algorithm, but a Bayesian Markov Chain Monte Carlo (MCMC) approach that provides as output the full posterior probability density (PPD) function, in a multi-dimensional form for all free parameters, $\vec{\lambda}^{cs}$, $\vec{\lambda}^{HX}$ and $\vec{\lambda}^{PX}$. The prior distribution is chosen to be uniform in the anisotropy parameters. Central values and confidence level (CL) intervals can be constructed from this multi-dimensional PPD by projections on the individual anisotropy parameters. This method ensures that correlations between the free parameters are taken into account and can be studied in detail. In addition, there are no convergence problems as observed in traditional fitting approaches, leading to a simple, stable, reliable and fast algorithm.

In summary, the framework used for the extraction of the polarization parameters requires these basic inputs:

- The muon 4-momentum vectors of the selected events in suitable windows around the pole mass (of the quarkonium state that is studied) in a prompt lifetime region, to minimize the contaminations of the data sample from any of the background sources.
- The fraction of inclusive background events in the selected sample, $f_{BG}$, including all background contributions, as evaluated from a fit to the dimuon mass distribution in case of the $\Upsilon(nS)$ analysis (see Sect. 4.2.3) and from fits to the dimuon mass and lifetime distributions in case of the $\psi(nS)$ analysis (see Sect. 4.3.3).
- The $(p_T, |y|, M_{\mu\mu}, \cos\vartheta, \varphi)$ distribution of the inclusive background events in the selected sample. In the analyses described in this thesis, the background model is given in a factorized form, as a 2-dimensional histogram of the $(\cos\vartheta, \varphi)$ background distribution, $A_{BG}(\cos\vartheta, \varphi)$, complemented by a 3-dimensional histogram of the $(p_T, |y|, M_{\mu\mu})$ background distribution, $K_{BG}(p_T, |y|, M_{\mu\mu})$. This implies the assumption that the $(\cos\vartheta, \varphi)$ background distribution does not change as a function of $p_T$, $|y|$ or $M_{\mu\mu}$, which is justified due to the small bin sizes used in the analyses. If the inclusive background constitutes a mixture of several background contributions with different kinematic and angular distributions, the background model histograms $A_{BG}$ and $K_{BG}$ describe the inclusive distributions, with the distributions of the individual contributions weighted accordingly. The $A_{BG}(\cos\vartheta, \varphi)$ distribution needs to be known for one frame only. In the specific analyses described here, the PX frame is chosen, as this is the frame with minimal $\cos\vartheta - \varphi$ correlations of acceptance and efficiency effects.
- The dimuon efficiencies $\epsilon_{\mu\mu}(\vec{p}^{\,\mu^+}, \vec{p}^{\,\mu^-})$, with $\vec{p}^{\,\mu^+}$ and $\vec{p}^{\,\mu^-}$ the 3-momentum vectors of the positive and negative muon, respectively.

The inputs mentioned above vary slightly between the individual analyses, and are therefore described in more detail in Sects. 4.2 and 4.3. This section is restricted to explain how these generic inputs are used to extract the polarization parameter results.

## *4.1.2  Background Subtraction*

The background subtraction algorithm removes a fraction $f_{BG}$ of events from the original data sample, which includes the signal plus the inclusive background contributions. Events that have a $(p_T, |y|, M_{\mu\mu}, \cos\vartheta, \varphi)$ distribution similar to the background model are selected with a likelihood-ratio method, and are removed from the sample before the determination of the polarization parameters. The algorithm is defined as follows. Similarly to $A_{BG}$ and $K_{BG}$, "signal plus background" histograms are built from the full data sample itself, $A_{S+BG}(\cos\vartheta, \varphi)$ and $K_{S+BG}(p_T, |y|, M_{\mu\mu})$. $A_{BG}$ and $K_{BG}$ are normalized to $f_{BG}$, while $A_{S+BG}$ and $K_{S+BG}$ are normalized to 1. For a given event, one calculates from the background model histograms $A_{BG}$

and $K_{BG}$ the likelihood $\mathcal{L}_{BG}$, under the background only hypothesis. Similarly, for the same event, one calculates the signal plus background likelihood $\mathcal{L}_{S+BG}$ from $A_{S+BG}$ and $K_{S+BG}$.

These likelihoods are used to classify the event with a likelihood ratio criterion, by drawing a random number $r$ from a uniform distribution, $r \in [0,1]$. If the condition $\mathcal{L}_{BG} > r \cdot \mathcal{L}_{S+BG}$ is fulfilled, the event is classified as "background-like" and removed from the sample. Due to the random nature of this background removal procedure, it has to be repeated several times ($n_{fit} = 50$, see Sect. 4.1.3), to smear out the fluctuations.

Figure 4.1 shows the performance of the background subtraction algorithm, illustrated by an example from the $\Upsilon(3S)$ data analysis, in one specific kinematical cell. The figures show the ratio of the distribution of the subtracted events with respect to the input background model distribution, as a function of $\cos \vartheta^{PX}$ (top left), $\varphi^{PX}$ (top right), dimuon $p_T$ (middle left), $|y|$ (middle right) and mass (bottom left), showing that the background subtraction procedure does not bias the kinematic and angular distributions. The bottom right panel shows the ratio of the fraction of subtracted events with respect to the input $f_{BG}$, for the $n_{fit} = 50$ extractions. The average is compatible with 1. Nevertheless, the spread shows the importance of repeating the background subtraction a sufficient number of times.

### 4.1.3 Posterior Probability Density of the Anisotropy Parameters

**Definition of the Parameter Likelihood**

After the background removal procedure, the remaining sample is considered background-free, and the determination of the anisotropy parameters starts from the remaining signal-only sample.

The "event-probability" $\mathcal{E}(\vec{p}^{\,\mu^+}, \vec{p}^{\,\mu^-}|\vec{\lambda})$ is a probability density function (PDF) of the lepton kinematics, defined as

$$\mathcal{E}(\vec{p}^{\,\mu^+}, \vec{p}^{\,\mu^-}|\vec{\lambda}) = \frac{1}{\mathcal{N}(\vec{\lambda})} \cdot W(\cos \vartheta, \varphi|\vec{\lambda}) \cdot \epsilon_{\mu\mu}(\vec{p}^{\,\mu^+}, \vec{p}^{\,\mu^-}) , \qquad (4.1)$$

where $\mathcal{N}(\vec{\lambda})$ is a normalization function, and $W(\cos \vartheta, \varphi|\vec{\lambda})$ is the assumed dilepton angular distribution (Eq. 2.5).

The event probability needs to be normalized in the 6-dimensional space of the lepton kinematics, $\vec{p}^{\,\mu^+}$ and $\vec{p}^{\,\mu^-}$. The normalization function $\mathcal{N}(\vec{\lambda})$ depends both on the specific value of $\vec{\lambda}$, and on the efficiency function $\epsilon_{\mu\mu}(\vec{p}^{\,\mu^+}, \vec{p}^{\,\mu^-})$,

$$\mathcal{N}(\vec{\lambda}) = \int \int W(\cos \vartheta, \varphi|\vec{\lambda}) \cdot \epsilon_{\mu\mu}(\vec{p}^{\,\mu^+}, \vec{p}^{\,\mu^-}) \mathrm{d}\vec{p}^{\,\mu^+} \mathrm{d}\vec{p}^{\,\mu^-} , \qquad (4.2)$$

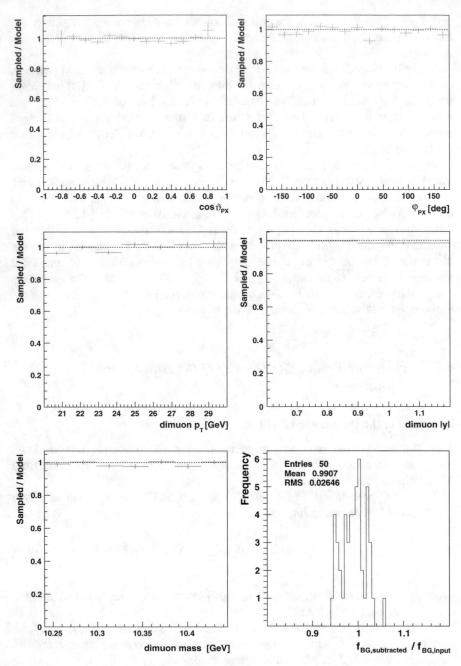

**Fig. 4.1** Performance of the background subtraction algorithm, in the kinematical bin $20 < p_T < 30\,\text{GeV}$, $0.6 < |y| < 1.2$ for the $\Upsilon(3S)$ data analysis. Ratio of subtracted event distribution with respect to the input background distribution, as a function of $\cos\vartheta^{PX}$ (*top left*), $\varphi^{PX}$ (*top right*), $p_T$ (*middle left*), $|y|$ (*middle right*) and mass (*bottom left*). Ratio of the fraction of subtracted events with respect to the input fraction of background events, for the $n_\text{fit} = 50$ extractions (*bottom right*)

and has to be recalculated for any change of $\vec{\lambda}$, which is the case in the MCMC method described below. In order to optimize the procedure, one can make use of the linearity of the dilepton angular distribution in $\vec{\lambda}$, and write the normalization as

$$\mathcal{N}(\vec{\lambda}) = \frac{1}{(3 + \lambda_\vartheta)} \cdot \left[ I_a + \lambda_\vartheta \cdot I_b + \lambda_\varphi \cdot I_c + \lambda_{\vartheta\varphi} \cdot I_d \right] , \qquad (4.3)$$

with $I_a$, $I_b$, $I_c$ and $I_d$ integrals of the type $\int \int \mathcal{F}(\cos\vartheta, \varphi)\mathrm{d}\cos\vartheta \, \mathrm{d}\varphi$ that can be calculated once in the beginning of the procedure, avoiding the need to recalculate the expensive integrals for each step of the MCMC. However, the integrals cannot be calculated analytically, given the complex structure of the efficiency. Therefore, a simple MC procedure is developed to approximate these integrals. Events are generated according to the efficiency and acceptance-corrected $(p_T, |y|, M_{\mu\mu})$ distributions $\mathcal{H}(p_T, |y|, M_{\mu\mu})$. The shape of $\mathcal{H}(p_T, |y|, M_{\mu\mu})$ is obtained by scanning over all data events, correcting for efficiency and acceptance. The acceptance is defined as the probability that for a dimuon event in a given $(p_T, |y|, M_{\mu\mu})$ cell the decay muons pass the single muon fiducial cuts on $|\eta^\mu|$ and $p_T^\mu$. The integration must be uniform over $\cos\vartheta$ and $\varphi$. This condition is realized by generating the decay distributions of the dimuon events according to a uniform distribution in $\cos\vartheta$ and $\varphi$. The data themselves cannot be used to calculate the integrals, because the dimuons present in the data are not, a priori, unpolarized, thus possibly biasing the calculation of the integrals.

This method to calculate the normalization integrals is unbiased and very fast, given that the integrals have to be evaluated only once, before the start of the MCMC.

**Sampling of the Parameter Likelihood**

The full likelihood is defined as

$$\mathcal{L}(\vec{\lambda}) = \prod_{i \in \{signal\}} \mathcal{E}(\vec{p}_i^{\,\mu^+}, \vec{p}_i^{\,\mu^-} | \vec{\lambda}) , \qquad (4.4)$$

with the index $i$ running over all signal events, after background subtraction. For practical reasons, the method discussed here relies on calculating the logarithm of the full likelihood defined in Eq. 4.4, $\log(\mathcal{L}(\vec{\lambda}))$. Instead of relying on a traditional maximization of the likelihood, it is preferable to employ a MCMC method to calculate the full log-likelihood distribution directly. The exploration of the $\vec{\lambda}$ parameter space could be done using a uniform sampling of points. However, this would be a very inefficient procedure, requiring an unreasonably large number of steps in the Markov Chain.

A more reasonable choice is the "Metropolis-Hastings" (MH) algorithm [17]. From a given point in parameter space, $\vec{\lambda}_j^{MH}$, with full log-likelihood $\log(\mathcal{L}(\vec{\lambda}_j^{MH}))$, one extracts a new set of the parameters, $\vec{\lambda}_{j+1}^{MH}$, from a given proposal probability distribution $\mathcal{P}_{MH}(\vec{\lambda}^{MH}, \vec{\sigma}^{MH})$, and recalculates the full log-likelihood $\log(\mathcal{L}(\vec{\lambda}_{j+1}^{MH}))$. If $\Delta = \log(\mathcal{L}(\vec{\lambda}_{j+1}^{MH})) - \log(\mathcal{L}(\vec{\lambda}_j^{MH})) > 0$, the new set of parameters is more likely than the previous one, and the new set is "accepted". In order to account for possible fluctuations, if this condition is not fulfilled, one draws a random number $r$ from a uniform distribution, $r \in [0,1]$. If $r < \exp(\Delta)$, the new parameter set is nevertheless accepted.

- In case the new parameter set is accepted by the MH algorithm, the Markov Chain continues with the next extraction starting from the new parameter set $\vec{\lambda}_{j+1}^{MH}$, and the parameter set $\vec{\lambda}_{j+1}^{MH}$ is written into the output n-tuple.
- In case the new parameter set is not accepted by the MH algorithm, the Markov Chain continues with the next extraction starting from the previous parameter set $\vec{\lambda}_{j}^{MH}$, and the parameter set $\vec{\lambda}_{j}^{MH}$ is written into the output n-tuple.

The proposal function $\mathcal{P}_{MH}(\vec{\lambda}^{MH}, \vec{\sigma}^{MH})$ of the MH algorithm is a Gaussian with mean $\vec{\lambda}^{MH}$ and predefined standard deviations $\vec{\sigma}^{MH} = (\sigma_\vartheta^{MH}, \sigma_\varphi^{MH}, \sigma_{\vartheta\varphi}^{MH})$, the "proposal widths". The performance (speed and efficiency) of the MH algorithm depends strongly on the chosen proposal widths. If the proposal widths are too large, the algorithm is inefficient, approximating in the limit of infinite $\vec{\sigma}^{MH}$ the uniform sampling of points. On the other hand, if the values chosen for $\vec{\sigma}^{MH}$ are too small, the Markov Chain does not have the freedom to quickly move towards the region of phase space where the distribution is centered. This problem is solved by defining a so-called "burn-in" period, with an arbitrary starting point of the Markov Chain and with relatively large values of $\vec{\sigma}^{MH}$, such that the center of the distribution is found quickly. After the burn-in period, the proposal widths $\vec{\sigma}^{MH}$ are adapted to correspond to the root mean square (RMS) of the 1-dimensional projections on each of the $\vec{\lambda}^{MH}$ parameters of the obtained burn-in distributions. The parameter values sampled during the burn-in period are not further used in the analysis. This procedure ensures that the center of the distribution is found quickly, at the same time decreasing the sensitivity on the initial starting point of $\vec{\lambda}^{MH}$.

Once the Markov Chain has proceeded over a pre-defined number of "samplings", the sampled points in the output n-tuple are an approximation of the posterior probability density function of the anisotropy parameters $\vec{\lambda}$, from which one can easily construct the PPD as a function of any combination of the parameters, as for example $\vec{\lambda}$. This procedure runs in parallel for each of the three reference frames, such that the PPD can be constructed independently for each of the frames of the measurement.

In the specific case of the CMS polarization analyses, the chosen starting point is $\vec{\lambda}^{MH} = (0, 0, 0)$ and $\vec{\sigma}^{MH} = (0.1, 0.1, 0.1)$. The number of burn-in extractions is chosen to be 10 k and the number of extractions after the burn-in is 50 k, large enough to ensure smooth distributions of the multi-dimensional PPD. Figure 4.2 (left) shows the 2-dimensional projections of the PPD on $\lambda_\vartheta$ and $\lambda_\varphi$ in the PX frame, for an example bin of the $\Upsilon(3S)$ data analysis, for various settings of the burn-in period, showing that the arbitrariness of the parameters of the burn-in period chosen for this framework has a negligible impact on the results.

Mostly due to the random nature of the background subtraction procedure defined in Sect. 4.1.2, but also due to the random nature of the likelihood sampling discussed here, it is advisable to repeat the full procedure, background subtraction

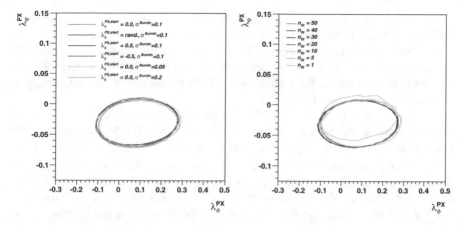

**Fig. 4.2** The individual lines show the 68.3% CL contours of the 2-dimensional projection of the PPD on $\lambda_\vartheta$ and $\lambda_\varphi$ in the PX frame, corresponding to the kinematical cell $|y| < 0.6$, $20 < p_T < 30$ GeV of the $\Upsilon(3S)$ data analysis, for various settings of the burn-in period (*left*), and after evaluating the contour for $n_{fit} =1, 5, 10, 20, 30, 40$ and $50$ repetitions (*right*)

and likelihood sampling, several times. Figure 4.2 (right) shows the 2-dimensional projections of the PPD on $\lambda_\vartheta$ and $\lambda_\varphi$ in the PX frame, after different numbers of repetitions of the procedure, $n_{fit}$, showing that the results are reasonably stable already after 30 repetitions. Therefore, for the analyses presented here, $n_{fit} = 50$ repetitions are employed. The PPDs of the $n_{fit}$ iterations are merged, building the "combined PPD", which then automatically includes in the spread of the PPD the uncertainties due to the effect of the random nature of the framework.

### 4.1.4 Extraction of the Results

#### Interpretation of the PPD

The value of highest posterior probability in its 1-dimensional projection is used as estimate of the central (most probable) value (MPV) of each of the polarization parameters, $\hat{\lambda}$. Given that the maximum of the resulting histograms can be affected by statistical fluctuations, the 1-dimensional PPD projections are fitted with a Gaussian in a small interval around the maximum.

The CLs at p% are estimated by finding the interval $[\hat{\lambda} - \vec{\sigma}_-, \hat{\lambda} + \vec{\sigma}_+]$, so that the intervals $[-\infty, \hat{\lambda} - \vec{\sigma}_-]$ and $[\hat{\lambda} + \vec{\sigma}_+, \infty]$ contain $(100 - p)/2\%$ of the integral over the 1-dimensional PPD each. The points $\hat{\lambda} - \vec{\sigma}_-$ and $\hat{\lambda} + \vec{\sigma}_+$ therefore correspond to the quantiles $q_{(1-p/100)/2}$ and $q_{1-(1-p/100)/2}$, respectively, of the 1-dimensional projection of the PPD. Symmetric uncertainties can be constructed by defining $\vec{\sigma}_s = (\vec{\sigma}_- + \vec{\sigma}_+)/2$ as the average of the asymmetric uncertainties.

An alternative definition of the CL regions would be the high-posterior-density (HPD) intervals, defined as the region that integrates p% of the distribution and whose end points have equal posterior probabilities. In case of symmetric PPDs, which is a good approximation of the results of the analyses in all cases, the definitions coincide. The former definition has the advantage that large polarizations can be excluded at a certain CL in a symmetric way.

The output of this method and the interpretation of the PPD is visualized in Fig. 4.3, showing results of an example bin from the $\Upsilon(3S)$ data analysis. The top two distributions show 1-dimensional projections of the combined PPD on $\lambda_\vartheta^{px}$ (left) and $\tilde{\lambda}$ for all frames (right), the middle row shows the $\lambda_\vartheta$-$\lambda_\varphi$ plane showing the 2-dimensional contours of the combined PPD for all frames (left) and the distribution of the most probable values of the individual $n_{\mathrm{fit}} = 50$ repetitions (right). The bottom row shows data distributions after background subtraction, compared to the angular distribution models corresponding to the best fit values of $\vec{\lambda}$, as a function of $\cos\vartheta^{px}$ (left) and $\varphi^{px}$ (right), including curves corresponding to the most extreme physical cases of modulations, indicating the level of sensitivity of the current data set.

In principle, the central value should be evaluated as the value that gives the highest posterior probability of the *full*-dimensional PPD. In cases of very broad and asymmetric PPD shapes, the simplification of evaluating the central value (and the corresponding CLs) from the 1-dimensional projections of the PPD can introduce a bias. This is for example the case for very low-$p_T$ regions, where the phase space coverage in $\cos\vartheta$ is very limited, leading to broad and asymmetric PPDs. However, in the regions accessed in the scope of the analyses described in this thesis, the corresponding effect was evaluated and found to be negligible.

**Treatment of Systematic Uncertainties**

Systematic uncertainties are directly incorporated into the PPD, whose spread, before this procedure, reflects statistical uncertainties only. For each source of systematic uncertainty, a probability distribution as a function of $\vec{\lambda}$ is defined, individually for each frame, kinematic cell and quarkonium state. These probability distributions describe the probability of the variations of the parameter value due to the individual source of systematic uncertainty. These distributions are in most cases assumed to be Gaussian, with a width corresponding to a shift of the anisotropy parameter due to a systematic variation given by the effect under study. An alternative possibility realized in the analyses is the definition of a uniform probability distribution, in a range that covers the possible variations of the anisotropy parameters at 100% CL, without assuming that any value is more probable than any other value within that range.

For each set of parameters in the output n-tuple of the PPD, a vector $\Delta\vec{\lambda}$ is constructed, with four parameters, including $\tilde{\lambda}$, individually for the output n-tuples of the three reference frames, by drawing random numbers from the defined probability distributions of the individual systematic effects. The random numbers of each effect are added to calculate $\Delta\vec{\lambda}$. The resulting vector is then added to the initial parameter vector of the entry of the n-tuple. This addition represents a smearing of the PPD,

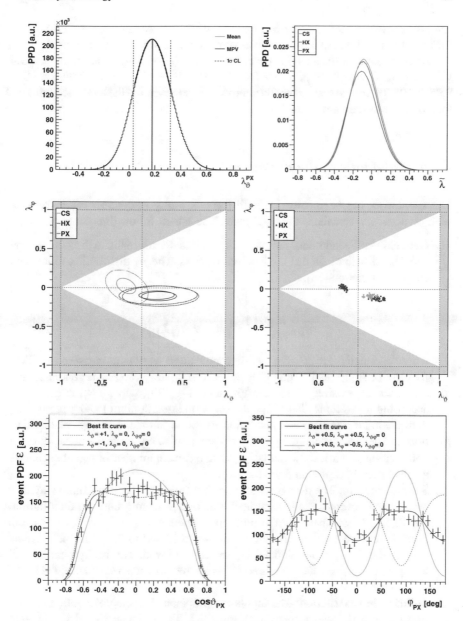

**Fig. 4.3** Results in the kinematic cell $20 < p_T < 30\,\text{GeV}$, $0.6 < |y| < 1.2$ for the $\Upsilon(3S)$. 1-dimensional projections of the combined PPD distribution of the parameter $\lambda_\vartheta^{PX}$ (*top left*) and $\tilde{\lambda}$ in all frames (*top right*). 2-dimensional projections (99.7 and 68.3% CL contours) of the combined PPD in the plane $\lambda_\vartheta$ versus $\lambda_\varphi$, shown for all frames (*middle left*). Distribution of most probable values of the $n_{\text{fit}} = 50$ extractions for parameters $\lambda_\vartheta$ and $\lambda_\varphi$ in all frames (*middle right*). Data distributions and fit results projected on the variables $\cos\vartheta$ (*bottom left*) and $\varphi$ (*bottom right*) in the PX frame. The *solid lines* represent the best-fit result, while the *dashed* and *dotted lines* correspond to the polarization hypotheses giving the maximum modulations

according to all sources of systematic uncertainty. This procedure is repeated 5 times, for all entries of the output n-tuple, to ensure a smooth smearing of the PPD.

For each result of the analyses described in this thesis, in addition to the central value and the CL at 68.3% of the statistical PPD only, the CLs are evaluated at 68.3, 95.5 and 99.7% for the systematically altered PPD representing the total uncertainties, including the systematic variations.

### 4.1.5   Validation of the Framework

The framework has been equipped with the possibility to conduct extensive tests with pseudo-data, so called "toy-MC" tests. These tests follow three steps.

1. **Generation**: Pseudo-data samples are generated with realistic dimuon $p_T$, $|y|$ and $M_{\mu\mu}$ distributions that are inspired by data. The $p_T$ distribution follows an empirical function [18],

$$ \frac{dN}{dp_T} \propto p_T \left[ 1 + \frac{1}{\beta - 2} \cdot \frac{p_T^2}{\langle p_T^2 \rangle} \right]^{-\beta}, \tag{4.5} $$

   whose parameters are fit to the $p_T$ differential cross sections as measured by CMS for the $\psi(nS)$ [19] and $\Upsilon(nS)$ [20] states. The rapidity variable is generated flat, supported by existing CMS measurements [19, 20], while $M_{\mu\mu}$ is generated according to a simple Gaussian centered at the quarkonium PDG masses [21], with a similar dimuon mass resolution as found in data. The angular distributions are generated according to any chosen ("injected") signal and background polarization scenarios, $\vec{\lambda}_S^{\text{inj.}}$ and $\vec{\lambda}_{BG}^{\text{inj.}}$, with a certain number of signal events $n_S$ and a chosen background fraction, $f_{BG}$.
2. **Reconstruction**: The detector response is simulated in a very simple way. For each pseudo-event a random number $r$ is drawn from a uniform distribution, $r \in [0,1]$, and compared to the dimuon efficiency. If $r < \epsilon_{\mu\mu}(\vec{p}^{\,\mu^+}, \vec{p}^{\,\mu^-})$, the event is removed. Otherwise, the event is considered to be "efficient". Events outside the single muon fiducial region are removed. The background model histograms $A_{BG}$ and $K_{BG}$ are filled with the "reconstructed" and accepted background pseudo-events.
3. **Polarization extraction**: The inputs are propagated to the polarization extraction framework, which determines the PPD. The efficiency used in this step (representing the assumptions made) can be chosen to be different than in the reconstruction-step (representing the "reality"), in order to estimate the effect of efficiency variations on the anisotropy parameters.

All tests performed to validate the framework and estimate systematic uncertainties with toy-MC studies rely on at least $n_{\text{toy}} = 50$ pseudo-experiments. For each pseudo-experiment the three steps above are conducted individually, resulting in

$n_{\text{toy}} = 50$ results for $\hat{\vec{\lambda}}$ and the corresponding statistical uncertainties. The median of the $\hat{\vec{\lambda}}$ results is denoted as "toy-MC result", while the median of the difference of the $\hat{\vec{\lambda}}$ results with respect to the injected polarization $\vec{\lambda}_S^{\text{inj.}}$ is quoted as "toy-MC bias". A bias is observed if the toy-MC bias is significantly different from 0.

The reliability of the error estimation can be assessed by studying the distribution of the standard scores $\vec{z}$ ("pull distributions"),

$$ \vec{z} = \frac{\hat{\vec{\lambda}} - \vec{\lambda}_S^{\text{inj.}}}{\vec{\sigma}_s} . \tag{4.6} $$

The pull distributions, for each component of $\vec{z}$, are expected to follow a standard normal distribution $G(0, 1)$. The distributions are fit with a Gaussian $G(\mu_p, \sigma_p)$. If $\mu_p$ is significantly different from 0, the framework is biased (corresponding to a toy-MC bias). If $\sigma_p$ is significantly different from 1, the uncertainties are biased. If $\sigma_p > 1$ ($\sigma_p < 1$) the uncertainties are underestimated (overestimated), on average.

Several toy-MC studies are conduced, for each of the individual quarkonium resonances, to validate the framework. Systematic uncertainties related to the framework itself are studied, factorized in three components:

- Uncertainties related to limited statistics are tested with toy-MC studies, using $n_S$ and $f_{BG}$ as estimated from data, with unpolarized signal and background.
- Uncertainties related to large signal polarizations are tested with toy-MC studies with $\lambda_\vartheta^{px} = \pm 0.5$. The mean of the absolute values of the toy-MC biases is used to define the associated systematic uncertainty.
- Uncertainties related to the background subtraction are tested with toy-MC studies using $f_{BG}$ as estimated from data, and three extreme background polarization scenarios, inspired by data (see test in Sect. 4.2.4). Again, the mean of the absolute values of the toy-MC biases is used to define the associated systematic uncertainty.

Figure 4.4 shows a subset of the results of the toy-MC studies as described above, by displaying the toy-MC biases of the individual studies for the parameters $\lambda_\vartheta$ (left) and $\lambda_\varphi$ (right) in the $\Upsilon(nS)$ analysis, in the frames where the largest effect is observed, in the region $|y| < 0.6$. Non-negligible toy-MC biases of these three tests are only observed outside of the kinematic region considered in this thesis, for $p_T < 10$ GeV, due to the low angular acceptance in this region. Nevertheless, systematic uncertainties are defined by adding the toy-MC biases of the three tests quadratically, resulting in uncertainties that are negligible with respect to the other sources of uncertainty (see Sects. 4.2.5 and 4.3.5).

Apart from the tests mentioned above, several other cross checks are conducted to validate the framework. The tests include several more signal and background polarization scenarios, including the signal polarizations as measured from data, and more extreme signal polarizations, e.g., $\lambda_\vartheta^{px} = \pm 1$. Furthermore, unreasonably large background fractions are successfully tested. Acceptance and efficiencies in all mentioned studies are taken to approximate realistic CMS conditions. The tests

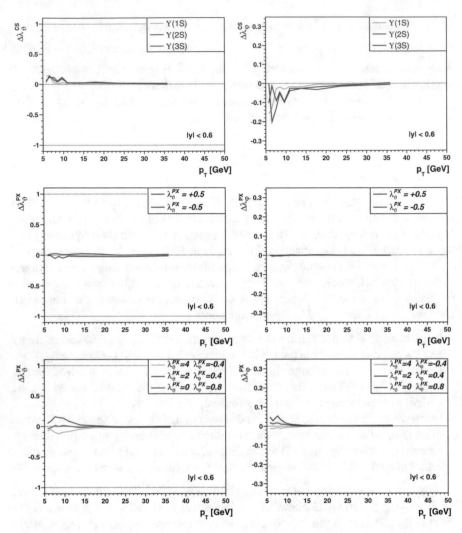

**Fig. 4.4** Toy-MC biases for $\lambda_\vartheta$ (*left*) and $\lambda_\varphi$ (*right*) in the region $|y| < 0.6$, as a function of $p_T$, for the studies using $n_S$ and $f_{BG}$ as estimated from data, for the three $\Upsilon(nS)$ states in the Collins–Soper frame (*top*), for the studies with various signal polarization scenarios in the PX frame (*middle*), and for studies with various background polarization scenarios, for the $\Upsilon(3S)$ analysis, in the PX frame (*bottom*)

cover all kinematic regions accessible by the analyses described in this thesis, and beyond.

Some of the tests show very small, but significant toy-MC biases. However, none of the tests have shown any non-negligible biases in the framework, comparable in size to other systematic uncertainties of the individual analyses, neither on the central value, nor on the uncertainties. Therefore, both the background subtraction procedure and the likelihood sampling algorithm can be regarded as validated.

The toy-MC framework is utilized for the determination of several systematic uncertainties, as discussed below.

## 4.2 Measurement of the $\Upsilon(nS)$ Polarizations

This section describes in detail the measurement of the $\Upsilon(nS)$ polarizations with the CMS detector in $pp$ collisions at $\sqrt{s} = 7\,\text{TeV}$, a core section of this thesis, given that I was the primary contributor to this analysis, which has paved the way for the corresponding $J/\psi$ and $\psi(2S)$ polarization analysis (Sect. 4.3). The results of this analysis have been published in Ref. [1].

### 4.2.1 $\Upsilon(nS)$ Data Processing and Event Selection

The analysis considers data collected at the CMS experiment in $pp$ collisions during the 2011 run. The total integrated luminosity used for this analysis is 4.9 fb$^{-1}$. The considered HLT paths, *HLT_DimuonX_Upsilon_Barrel* (with X = 5, 7 or 9), and the corresponding L1 seed, *L1_DoubleMu0_HighQ*, were discussed in detail in Sect. 3.2.4. They were mostly unchanged throughout the run, except for the minimum dimuon $p_T$ requirement, which was adapted depending on the instantaneous luminosity at the beginning of the LHC fill.

The triggered events are reconstructed with standard CMS software, requiring that the two reconstructed opposite-sign muons are matched to the two muons that "fired" the HLT path, to ensure correct treatment of the muon trigger efficiencies. This matching criterion is imposed for both the physics data analysis, as well as for the efficiency studies. The muons are subjected to a standard set of muon selection cuts, including cuts on the number of hits in the silicon pixel detectors, a cut on the number of hits in the silicon strip detectors, a cut on the reduced $\chi^2$ of the track fit and geometrical cuts, ensuring that the muons originate from the BS. The two muons are required to originate from the same vertex, with a vertex $\chi^2$ probability $P_{vtx}^{\mu\mu}$ larger than 1%. Furthermore, cowboy dimuons are rejected (see Sect. 3.2.4), and a cut on the lifetime significance $|\ell/\sigma_\ell^{PE}| < 2$ is applied (with $\sigma_\ell^{PE}$ the "per-event" uncertainty on $\ell$), reducing the continuum background by around 25%, while retaining around 95% of the prompt signal.

As will be discussed in the following Sect. 4.2.2, the muon efficiencies can only be reliably determined for sufficiently high values of $p_T^\mu$, depending on $|\eta^\mu|$,

$$|\eta^\mu| < 1.2 : \quad p_T^{\mu,\min} = 4.5\,\text{GeV} ,$$
$$1.2 < |\eta^\mu| < 1.4 : \quad p_T^{\mu,\min} = 3.5\,\text{GeV} , \qquad (4.7)$$
$$1.4 < |\eta^\mu| < 1.6 : \quad p_T^{\mu,\min} = 3\,\text{GeV} .$$

This region is denoted as the "single muon fiducial region". This requirement ensures that the events used in the analysis are not dangerously close to the acceptance edges, where the efficiency cannot be reliably determined. On the other hand, these single muon fiducial cuts restrict the angular phase space, especially in $\cos\vartheta^{px}$, for low-$p_T$ dimuons, reducing the accuracy of the measurement of the polarization parameters. These cuts constitute a reasonable compromise between decreasing the systematic uncertainties related to the muon efficiencies, and an increase in statistical uncertainty due to the reduced angular acceptance and the number of rejected events. The angular acceptance is visualized in Fig. 4.5, showing the data distributions in $\cos\vartheta$ and $\varphi$ in the PX, HX and Collins-Soper frames, in the $\Upsilon(1S)$ mass region, for several $p_T$ ranges. This figure includes a very low $p_T$ bin not used in the analysis, to illustrate the magnitude of the effect towards lower $p_T$.

The reduced angular acceptance is the main reason to restrict the analysis to dimuon $p_T > 10$ GeV. The analysis is conducted in two ranges of dimuon rapidity, $|y| < 0.6$ and $0.6 < |y| < 1.2$, in 5 bins of dimuon $p_T$, the bins being defined by the borders $(10, 12, 16, 20, 30, 50)$ GeV. The bins are defined to be as narrow as possible while retaining enough signal events to ensure a reasonable measurement of the polarization parameters.

To allow for a detailed study of certain effects, high-statistics signal MC samples are produced, for each $\Upsilon(nS)$ state, with a simple particle-gun approach. This approach is considerably less time-consuming than more sophisticated approaches such as for example PYTHIA [22]. The angular distribution is generated isotropically, corresponding to the polarization parameters $\vec{\lambda} = (0, 0, 0)$. The rapidity dimension is generated flat, while the $p_T$ distribution is inspired by the empirical function defined in Eq. 4.5, with different parameters for each $\Upsilon(nS)$ state. The samples are subjected to the full detector simulation, including the time-dependent trigger configuration, and include the effect of final state radiation (FSR) due to the QED radiation of the final state muons.

### 4.2.2  $\Upsilon(nS)$ Efficiencies

The reliable and accurate determination, validation and parametrization of the dimuon efficiency $\epsilon_{\mu\mu}(\vec{p}^{\,\mu^+}, \vec{p}^{\,\mu^-})$ is a vital necessity for a successful measurement of quarkonium polarization. This study has been conducted with a great effort, with major contributions from Ilse Krätschmer, Hermine Wöhri, Linlin Zhang and myself, and documented in detail in the CMS analysis notes [4, 5] and in Ref. [23]. The dimuon efficiency,

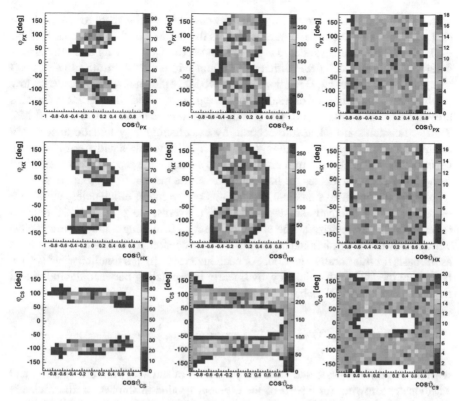

**Fig. 4.5** 2-dimensional angular distributions of $\cos\vartheta$ and $\varphi$ in the PX (*top*), HX (*middle*) and Collins–Soper (*bottom*) frames, after all selection cuts, projected from $1\sigma$ windows around the $\Upsilon(1S)$ pole masses, in the dimuon rapidity region $|y| < 0.6$ and in three dimuon $p_T$ ranges: $5 < p_T < 6$ GeV (*left*), $10 < p_T < 12$ GeV (*middle*), and $30 < p_T < 50$ GeV (*right*)

$$\epsilon_{\mu\mu}(\vec{p}^{\,\mu^+}, \vec{p}^{\,\mu^-}) = \epsilon_{\mu^+}(\vec{p}^{\,\mu^+}) \cdot \epsilon_{\mu^-}(\vec{p}^{\,\mu^-}) \cdot \rho(\vec{p}^{\,\mu^+}, \vec{p}^{\,\mu^-}) \cdot \epsilon_{vtx}(\vec{p}^{\,\mu^+}, \vec{p}^{\,\mu^-}) \,, \quad (4.8)$$

can be factorized in the efficiencies of the two individual muons, $\epsilon_{\mu^+}$ and $\epsilon_{\mu^-}$, and has to be corrected for effects originating from the presence of a second muon, referred to as the "$\rho$-factor", and an inconspicuous factor $\epsilon_{vtx}$, describing the inefficiency due to the cut on $P_{vtx}^{\mu\mu}$. For the measurement of the polarization, the normalization of the efficiencies is irrelevant, as only differences in shape (as a function of the muon momenta and angular variables) can affect the polarization measurement.

**Muon Efficiencies**

The single muon efficiency $\epsilon_\mu$ (which is symmetric in the muon charge) can be factorized into the muon trigger efficiencies ("L1·L2" and "L3") and the muon reconstruction efficiencies ("muon tracking", "muon identification" and "muon quality"). The individual factorized parts of the muon efficiencies are studied with the $T\&P$ method, as a function of $p_T^\mu$ and $|\eta^\mu|$, from a data sample of J/$\psi$ events collected with dedicated efficiency triggers, which are unbiased with respect to the presence of a second muon. The full muon efficiency is the product of the individual factors.

The $T\&P$ method is validated with a MC procedure, comparing MC efficiencies determined with ("MC $T\&P$ efficiencies") and without ("MC truth efficiencies") using the $T\&P$ approach. Deviations of the two MC efficiencies are observed at low $p_T^\mu$, leading to the definition of a safe fiducial region, for which the $T\&P$ efficiencies are validated, as shown in Eq. 4.7. Residual differences of the MC truth and MC $T\&P$ efficiencies are taken into account by establishing a systematic uncertainty related to the $T\&P$ approach. This systematic effect is evaluated with toy-MC studies (Sect. 4.1.5), by using MC truth efficiencies to simulate the reconstruction step and using MC $T\&P$ efficiencies for the extraction of the polarization parameters.

Not to fall victim to statistical fluctuations of the muon efficiencies, which are evaluated from a relatively small sample of $J/\psi$ events, the $p_T^\mu$ shape of the muon efficiencies is parametrized. The shape of the MC truth single muon efficiencies, which are evaluated with much higher statistical precision, in finer bins in $p_T^\mu$, is used as the basis for this parametrization. For each interval in $|\eta^\mu|$, a function $\epsilon_\mu^{\text{MCtruth}}(p_T^\mu)$ is built from the MC truth efficiency, by linearly interpolating between the individual measured points. A PDF is built,

$$\epsilon_\mu^{\text{DataT\&P}}(p_T^\mu) = \epsilon_\mu^{\text{MCtruth}}\left(\frac{p_T^\mu}{p_{T_{\text{scale}}}} - p_{T_{\text{shift}}}\right) + \epsilon_{\text{shift}}\,, \tag{4.9}$$

with the freedom to scale and shift the $p_T^\mu$ values of the MC truth efficiencies, and additionally allowing for a shift of the efficiency value of the MC truth efficiency function. A $\chi^2$ function is built, connecting the PDF with the data $T\&P$ efficiency results, as a function of three free parameters $p_{T_{\text{scale}}}$, $p_{T_{\text{shift}}}$ and $\epsilon_{\text{shift}}$. The $\chi^2$ function is minimized by a MINUIT [14] implementation.

Additionally to the central curve, representing the main parametrization of the data $T\&P$ efficiencies, curves representing the uncertainty on the central curve are provided. From the full information of the covariance matrix of the fit, the $3 \times 3$ matrix of the eigenvectors $O_{ij}$ is constructed. This matrix is multiplied with the matrix $L_{jk}$, representing the diagonal matrix containing the square root of the eigenvalues. The shift matrix $S_{ik} = O_{ij} \cdot L_{jk}$ represents the shift of the free fit parameters in their respective eigenbasis (their eigenvalues), rotated into the parameter basis. The shift matrix $S_{ik}$ contains one shift vector for each free fit parameter. This shift can be added or subtracted to the central values of the fit parameters, resulting in two error curves for each of the three free fit parameters, corresponding to a "positive" ("negative") variation each. This procedure ensures that the error band is correctly evaluated, taking into account the correlations between the parameters. Figure 4.6 shows the parametrization in three example $|\eta^\mu|$ bins, together with the error curves.

These error curves are used to establish systematic uncertainties on the parametrization. These effects are evaluated with toy-MC experiments, using in the reconstruction step the central parametrization, and in the extraction of the polarization the individual error curves of the parametrization. These toy-MC studies are performed

**Fig. 4.6** Central
parametrization (*red line*) of
the $p_T^{\mu}$ shape of the data
*T & P* efficiencies (*black
markers*), in three example
bins of $|\eta^{\mu}|$, and the
corresponding error curves
(*dotted blue lines*)

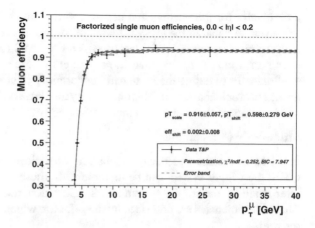

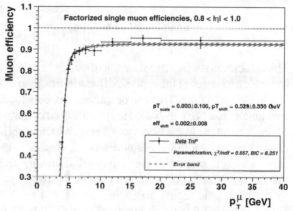

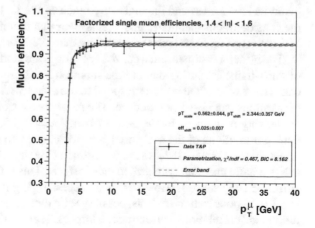

individually for the positive and negative variations for each parameter. The average absolute value of the toy-MC biases of the studies corresponding to the positive and negative variations is further used. The toy-MC bias averages corresponding to the error curves of the three fit parameters of the parametrization are summed quadratically to define the systematic uncertainty related to the parametrization of the muon efficiencies itself. This uncertainty also covers those related to the statistical uncertainties of the data-driven muon efficiency evaluation.

**Muon-Pair Correlations**

In cases where the two decay muons are very close-by in the muon detectors at CMS, the trigger system can be inefficient. In these situations, the dimuon can be interpreted by the trigger system as a single muon, and thus, the event is rejected. These inefficiencies are described by the $\rho$-factor, which is calculated from MC truth quantities,

$$\rho(\vec{p}^{\,\mu^+}, \vec{p}^{\,\mu^-}) = \frac{\epsilon_{\mu\mu}^{\text{MCtruth}}(\vec{p}^{\,\mu^+}, \vec{p}^{\,\mu^-})}{\epsilon_{\mu}^{\text{MCtruth}}(\vec{p}^{\,\mu^+}) \cdot \epsilon_{\mu}^{\text{MCtruth}}(\vec{p}^{\,\mu^-})} . \tag{4.10}$$

The $\rho$-factor is studied as a function of $\cos \vartheta^{PX}$ and $\varphi^{PX}$, in fine bins of dimuon $p_{\mathrm{T}}$,[1] in the rapidity ranges of the analysis. These inefficiencies are prominent for high-$p_{\mathrm{T}}$ and low-mass quarkonia. This can be understood considering that the two-body decay in two muons is a back-to-back decay in the quarkonium rest frame. The opening angle of the two muons in the laboratory frame, which is highly correlated with the distance of the two muons in the muon detectors, depends on the boost of the quarkonium state, which is proportional to the quarkonium $p_{\mathrm{T}}$ and inversely proportional to the quarkonium mass. The studies show that the muon-pair correlation effects are small in the kinematic region accessible with the $\Upsilon(nS)$ data used in this analysis. As shown in Sect. 4.3.2, this is not the case in the $\psi(nS)$ analysis, especially for the $\mathrm{J}/\psi$, given its lower mass and the higher $p_{\mathrm{T}}$ reach of the CMS data for this state. Given the considerations mentioned above, it is clear that muon-pair correlations affect the $\mathrm{J}/\psi$ data starting from lower values of $p_{\mathrm{T}}$ than the $\Upsilon(nS)$ data.

The $\rho$-factor 2-dimensional $\cos \vartheta^{PX}$-$\varphi^{PX}$ maps are studied by evaluating their level of "non-flatness", fitting them to the function Eq. 2.5, individually for each $p_{\mathrm{T}}$-$|y|$ cell. The results of these fits correspond to the effective $\vec{\lambda}$ that the residual muon-pair correlations can possibly introduce. The results vary from bin to bin, the maximum "polarization" induced by the $\rho$-factor being $\vec{\lambda} = (0.05, 0.01, 0.01)$. Given the statistical limitations of the MC based $\rho$-factor maps, which could introduce artificial effects in the analysis, the $\rho$-factor is assumed to be flat, and the worst case $\vec{\lambda}$ is taken as systematic uncertainty related to these effects. This is the optimal approach for the current $\Upsilon(nS)$ analysis. However, when accessing higher-$p_{\mathrm{T}}$ regions in a forthcoming $\Upsilon(nS)$ polarization analysis, possibly with data collected in 2012 or beyond, the muon-pair correlations will introduce larger effects, and will have to be taken into

---

[1]Binning of the $\rho$-factor studies: 10 bins in $\cos \vartheta^{PX}$, 12 bins in $\varphi^{PX}$ and 5 bins in dimuon $p_{\mathrm{T}}$, covering the range $10 < p_{\mathrm{T}} < 50\,\text{GeV}$.

account, as has been done already in the analysis of the $\psi(nS)$ polarizations with data collected in 2011, as shown in Sect. 4.3.2.

A different dimuon effect on the efficiencies can originate from the requirement on the dimuon vertex $\chi^2$ probability, quantified by evaluating $\epsilon_{vtx}$. The $\cos\vartheta^{PX}$-$\varphi^{PX}$ dependence of this efficiency is evaluated both with data $T\&P$ methods, as well as from MC. The angular distribution of $\epsilon_{vtx}$ is rather flat. The effect is quantified by conducting toy-MC experiments where the angular shape of $\epsilon_{vtx}$, as evaluated by MC, is used in the reconstruction step of the pseudo experiments, while for the polarization extraction step $\epsilon_{vtx}$ is assumed to be flat. Given the statistical limitations of the evaluation of the data-driven $\epsilon_{vtx}$, it is advantageous to assume it to be flat in the data analysis. The toy-MC bias of the study is completely negligible, therefore no systematic uncertainty is assigned.

A vital cross check of the evaluation and treatment of the dimuon efficiencies, as well as the framework itself, is the so-called "MC-closure test". The polarization parameters of the unpolarized particle-gun $\Upsilon(nS)$ signal MC samples are extracted with the method described in this section, without background subtraction. The results of $\vec{\lambda}$ are compatible with 0, in all kinematical bins, frames and for all $\Upsilon(nS)$ states, considering the statistical uncertainties and the relevant systematic uncertainties related to the efficiencies and the framework. This test validates the strategy of the dimuon efficiency treatment as well as, once again, the likelihood sampling discussed above.

### 4.2.3 $\Upsilon(nS)$ Mass Distribution

One essential feature of any quarkonium polarization analysis is the understanding of the dimuon mass distribution, $M_{\mu\mu}$. This is required to ensure reasonable definitions of regions around the signal peaks, characterized by increased signal purities, and background enriched regions far away from the signal peaks, that can be used to devise the background model distributions. Furthermore, from the information about the mass distribution, one can calculate the fraction of background events in the signal-enriched regions.

The $\Upsilon(nS)$ signal mass shapes are reasonably well modeled by one Crystal-Ball PDF for each state. The CB function is characterized by a Gaussian core, describing the detector resolution, and a power-law tail towards lower masses, describing the FSR tail, shifting the reconstructed dimuon masses towards smaller values. As the natural widths of the $\Upsilon(nS)$ states [21] are negligible with respect to the dimuon mass resolution, one can abstain from convoluting the detector response with a Breit-Wigner PDF, and the signal distributions are faithfully described by the CB functions. The Gaussian core is described by two parameters, $\mu_\Upsilon$ and $\sigma_\Upsilon$. The tail is described by the parameters $\alpha_\Upsilon^{CB}$ and $n_\Upsilon^{CB}$. The CB function is equivalent to a Gaussian for $M_{\mu\mu} > \mu_\Upsilon - \alpha_\Upsilon^{CB} \cdot \sigma_\Upsilon$, while for lower masses the CB function constitutes a power-law function with power $n_\Upsilon^{CB}$. The $\alpha_\Upsilon^{CB}$ and $n_\Upsilon^{CB}$ are free fit parameters, assumed to be identical for all $\Upsilon(nS)$ states. The parameters $\mu_\Upsilon$ and $\sigma_\Upsilon$ are only free for the $\Upsilon(1S)$ state, while the $\mu_\Upsilon$ parameters of the $\Upsilon(2S)$ ($\Upsilon(3S)$) are fixed by the difference of the

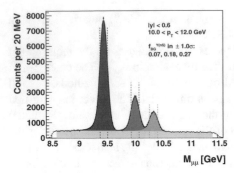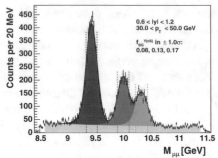

**Fig. 4.7** Dimuon mass distributions and fit results in the $\Upsilon$ mass region, for rapidity regions $|y| < 0.6$ at low $p_T$ (*left*) and $0.6 < |y| < 1.2$ at high $p_T$ (*right*), visualizing the individual contributions, and the definition of the individual $\Upsilon(nS)$ mass signal regions

masses $M_{\Upsilon(2S)}^{PDG}$ ($M_{\Upsilon(3S)}^{PDG}$) with respect to $M_{\Upsilon(1S)}^{PDG}$ [21]. The $\sigma_{\Upsilon}$ parameters of the heavier states are fixed by assuming that the dimuon mass resolution increases linearly with the dimuon mass, $\sigma_{\Upsilon(nS)} = \sigma_{\Upsilon(1S)} \cdot M_{\Upsilon(nS)}^{PDG}/M_{\Upsilon(1S)}^{PDG}$.

The $\mu\mu$ continuum background is suitably described with a polynomial of $2^{nd}$ order, whose parameters are fixed by an initial binned ML fit excluding the region $M_{\mu\mu} \in [8.9, 10.6]\,\text{GeV}$, ensuring that the fit is not biased by the presence of signal contributions. In a second step, the parameters of the $\Upsilon(nS)$ signal shapes, as well as their normalizations, are estimated from a binned ML fit to the full mass region $M_{\mu\mu} \in [8.6, 11.4]\,\text{GeV}$, slightly tighter than the trigger mass window, avoiding trigger resolution effects biasing the mass distribution.

The chosen models ensure a suitable description of the $\Upsilon(nS)$ mass spectrum, as can be appreciated by Fig. 4.7, showing the data distributions and the fit results for two example bins. The mass resolution at $M_{\Upsilon(1S)}^{PDG}$ is better than 70 MeV in the mid-rapidity bin, allowing a good separation of the $\Upsilon(2S)$ and $\Upsilon(3S)$ peaks in the analysis. However, in the forward-rapidity bin the mass resolution is around 95 MeV, resulting in a considerable overlap of the $\Upsilon(2S)$ and $\Upsilon(3S)$ peaks.

The information from the $\Upsilon$ mass fit allows us to define five $\Upsilon$ mass regions (in GeV) as

Left sideband ($\Upsilon$ LSB):   $[8.6, \; \mu_{\Upsilon(1S)} - n_{\sigma_\Upsilon}^{LSB} \cdot \sigma_{\Upsilon(1S)}]$ ,

Signal regions ($\Upsilon(nS)$ SR):   $[\mu_{\Upsilon(nS)} - n_{\sigma_\Upsilon} \cdot \sigma_{\Upsilon(nS)}, \; \mu_{\Upsilon(nS)} + n_{\sigma_\Upsilon} \cdot \sigma_{\Upsilon(nS)}]$ ,   (4.11)

Right sideband ($\Upsilon$ RSB):   $[\mu_{\Upsilon(3S)} + n_{\sigma_\Upsilon}^{RSB} \cdot \sigma_{\Upsilon(3S)}, \; 11.4]$ ,

with $n_{\sigma_\Upsilon}^{LSB} = 4.0, n_{\sigma_\Upsilon} = 1$ and $n_{\sigma_\Upsilon}^{RSB} = 3.5$. The value for $n_{\sigma_\Upsilon}^{LSB}$ is chosen to be larger than $n_{\sigma_\Upsilon}^{RSB}$ due to the FSR tail of the $\Upsilon(1S)$. The chosen value of $n_{\sigma_\Upsilon} = 1$, smaller than intuition might dictate, is motivated further below. Figure 4.8 (left) shows the background fractions $f_{BG}$ in the individual $\Upsilon(nS)$ SRs, evaluated by integration of the signal and background mass PDFs in the $\Upsilon(nS)$ SRs. The background fraction is largest in the low-$p_T$ bins, and – mostly due to the worse dimuon mass resolution

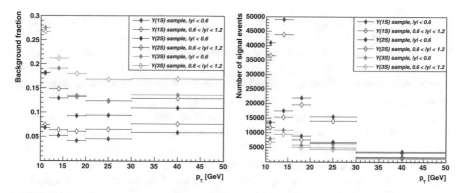

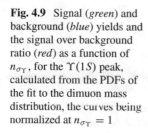

**Fig. 4.8** Background fraction $f_{BG}$ (*left*) and estimated number of $\Upsilon(nS)$ signal events (*right*) in the $\Upsilon(nS)$ SRs, as a function of $p_T$ for both rapidity ranges

**Fig. 4.9** Signal (*green*) and background (*blue*) yields and the signal over background ratio (*red*) as a function of $n_{\sigma_\Upsilon}$, for the $\Upsilon(1S)$ peak, calculated from the PDFs of the fit to the dimuon mass distribution, the curves being normalized at $n_{\sigma_\Upsilon} = 1$

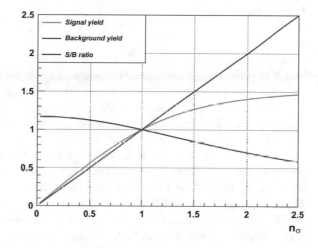

– larger in the forward-rapidity bin, ranging from around 27% for the $\Upsilon(3S)$ in the worst case to only around 4% for the $\Upsilon(1S)$ in the best case. The estimated number of $\Upsilon(nS)$ signal events in the $\Upsilon(nS)$ SRs is shown in the right panel of Fig. 4.8, ranging from around 50 k $\Upsilon(1S)$ signal events at low $p_T$ to around 1 k $\Upsilon(3S)$ signal events at high $p_T$.

The value defining the widths of the signal regions, $n_{\sigma_\Upsilon} = 1$, is chosen to be rather small with respect to comparable analyses, where a value of $n_{\sigma_\Upsilon} = 2.5$ or 3 would be regarded as "standard choice". The main reason for this choice is to reduce the background fractions in the signal regions, as the background is the source of the largest systematic uncertainty in this analysis (see Sect. 4.2.5), which is found to scale almost in a linear way with the background fraction. The pedagogical Fig. 4.9 shows the $n_{\sigma_\Upsilon}$-dependence of the signal and background yields, normalized at $n_{\sigma_\Upsilon}=1$, as well as the signal over background ratio, calculated from the output of the mass fit. If, instead of $n_{\sigma_\Upsilon} = 1$, the analysis would use a mass window of $n_{\sigma_\Upsilon} = 2.5$, the signal yield would increase by around 45%, improving the statistical uncertainty by around 20%. At the same time the background yield would be increased by a factor of 2.5, increasing $f_{BG}$ (and therefore the systematic associated uncertainty) by around 70%. Moreover, the choice of $n_{\sigma_\Upsilon} = 1$ is motivated by keeping the signal cross-feed fractions of the $\Upsilon(2S)$ and $\Upsilon(3S)$ states in the forward-rapidity bin reasonably small. With this choice, the fraction of $\Upsilon(3S)$ ($\Upsilon(2S)$) signal events in the $\Upsilon(2S)$ SR ($\Upsilon(3S)$ SR) is 4% (3%) in the worst cases. This contamination is neglected in this analysis, which is justified a posteriori, given that no significant differences between the polarizations of the $\Upsilon(nS)$ states are observed (see Sect. 4.2.6).

However, the tight dimuon mass window leads to the necessity of a (small) bias correction. Studies on MC show that the decay angular distribution is affected by tight cuts on the dimuon mass, because this distorts the distribution of the opening angle of the two decay muons, which affects the $\cos\vartheta\text{-}\varphi$ distribution, biasing the extracted polarization parameters. Therefore, the resulting bias of the angular distributions has to be corrected for,

$$\vec{\lambda}' = \vec{\lambda}^{n_{\sigma_\Upsilon}} + \Delta\vec{\lambda}^{n_{\sigma_\Upsilon}} , \qquad (4.12)$$

with $\vec{\lambda}^{n_{\sigma_\Upsilon}}$ the measured polarization parameters for a certain dimuon mass window, and $\vec{\lambda}'$ the corrected polarization parameters. The vector $\Delta\vec{\lambda}^{n_{\sigma_\Upsilon}}$ is the correction of the bias, which is applied by altering the PPD through shifting the entries of the output n-tuple by $\Delta\vec{\lambda}^{n_{\sigma_\Upsilon}}$. It is obtained by comparing the polarization parameters measured from the signal MC samples with and without a cut on the dimuon mass. The corrections are found to be identical for the three $\Upsilon(nS)$ states, so that they can be combined, to increase the statistical accuracy of these MC corrections. The top panels of Fig. 4.10 show the corrections $\Delta\vec{\lambda}^{n_{\sigma_\Upsilon}}$ for the parameters that are affected the most, $\lambda_\varphi^{PX}$ and $\lambda_{\vartheta\varphi}^{PX}$, in the mid-rapidity range, for the cases $n_{\sigma_\Upsilon} = 1$ and $n_{\sigma_\Upsilon} = 3$. These corrections are negligible in the case of $\lambda_\vartheta^{PX}$.

The MC corrections are tested by repeating the full data analysis with $n_{\sigma_\Upsilon} = 3$, where the correction is negligible, and comparing the results of the corrected analyses with the different choices $n_{\sigma_\Upsilon} = 1$ and $n_{\sigma_\Upsilon} = 3$. The observed results are compatible, as is shown in the bottom two rows of Fig. 4.10, therefore validating the MC corrections, and consequently no systematic uncertainty is associated to these MC corrections.

In summary, the choice of $n_{\sigma_\Upsilon} = 1$ is preferred with respect to a "usual choice" of a larger $n_{\sigma_\Upsilon}$ despite the necessity of the MC corrections and at the cost of rejecting around 1/3 of the signal, because of the reduced systematic uncertainty associated with the background model, and the reduced signal cross-feed.

A similar effect is found in the $\psi(nS)$ analysis regarding cuts on the lifetime variable $\ell$ (see Sect. 4.3.3). A possible effect of the selection cut $|\ell/\sigma_\ell^{PE}| < 2$ in the $\Upsilon(nS)$ analysis was studied by comparing results on data with and without this cut. No significant effect was found in the $\Upsilon(nS)$ analysis.

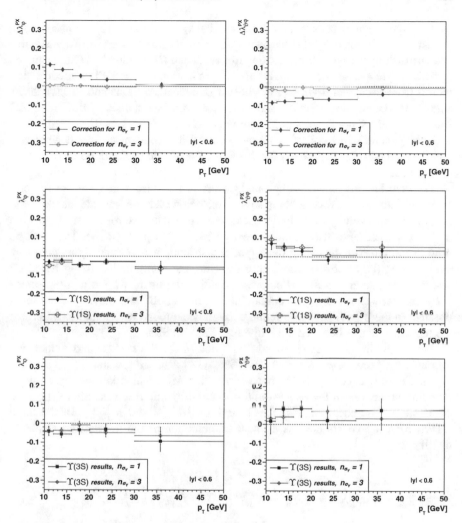

**Fig. 4.10** Corrections (*top*) $\Delta\vec{\lambda}^{n_{\sigma\Upsilon}}$ for $n_{\sigma\Upsilon} = 1$ (*blue*) and $n_{\sigma\Upsilon} = 3$ (*green*), for the parameters $\lambda_{\varphi}^{PX}$ (*left*) and $\lambda_{\vartheta\varphi}^{PX}$ (*right*) for the range $|y| < 0.6$. Measured parameters from $\Upsilon(1S)$ (*middle*) and $\Upsilon(3S)$ (*bottom*) data after the MC correction for different choices of the mass window $n_{\sigma\Upsilon} = 1$ (*black*) and $n_{\sigma\Upsilon} = 3$ (*red*), showing statistical uncertainties only

## 4.2.4 Determination of the $\Upsilon(nS)$ Background Model

With the definition of the mass signal regions, the estimation of the background fractions $f_{BG}$, and the definition of the $\Upsilon$ efficiencies, three out of the four inputs listed in Sect. 4.1.1 are fully defined. The remaining inputs to be defined are the background model histograms $A_{BG}(\cos\vartheta, \varphi)$ and $K_{BG}(p_T, |y|, M_{\mu\mu})$, which are discussed here.

The $\mu\mu$ continuum background is the only background source to be considered in this analysis. The majority of these background events are dimuons from open charm and open beauty decays, i.e., originating from $D\bar{D}$ and $B\bar{B}$ production. Other sources contributing to a lesser extent, are Drell-Yan production [24] as well as uncorrelated muons from decays-in-flight from light hadrons (charged pions and kaons). Their angular and kinematic distributions could in principle be obtained by dedicated MC simulations of the dominantly contributing background processes. However, in line with the general CMS policy to follow a model-independent approach whenever possible, a data-driven approach for the estimation of the background distributions is followed.

The distributions of the background events in the $\Upsilon(nS)$ SRs are not directly accessible, due to the overwhelming signal yield. However, the mass sidebands (SBs) provide the opportunity to study the background distributions, for masses below the $\Upsilon(1S)$ peak (LSB), and for masses beyond the $\Upsilon(3S)$ peak (RSB). They can be used to approximate the background model distributions under the signal peaks by interpolating the SB distributions into the SRs, the interpolation being based on assumptions that are discussed below. The definition of the SBs is a compromise between decreasing the signal contamination in the SBs and staying as close as possible to the $\Upsilon(nS)$ peaks, to retain enough background events and to minimize the mass differences between the individual SBs and SRs.

The events in the mass sidebands are filled in the SB background model histograms, $A_{BG}^{LSB}(\cos\vartheta, \varphi)$, $K_{BG}^{LSB}(p_T, |y|, M_{\mu\mu})$, $A_{BG}^{RSB}(\cos\vartheta, \varphi)$ and $K_{BG}^{RSB}(p_T, |y|, M_{\mu\mu})$. The only exception is the mass dimension $M_{\mu\mu}$, which is filled by drawing random numbers from the background mass PDF within the mass SR of the state under study. The background model histograms $A_{BG}^{\Upsilon(nS)}$ and $K_{BG}^{\Upsilon(nS)}$, for the three $\Upsilon(nS)$ SRs, are constructed as a linear combination of the normalized sideband distributions,

$$A_{BG}^{\Upsilon(nS)} = f_{LSB}^{\Upsilon(nS)} \cdot A_{BG}^{LSB} + (1 - f_{LSB}^{\Upsilon(nS)}) \cdot A_{BG}^{RSB},$$
$$K_{BG}^{\Upsilon(nS)} = f_{LSB}^{\Upsilon(nS)} \cdot K_{BG}^{LSB} + (1 - f_{LSB}^{\Upsilon(nS)}) \cdot K_{BG}^{RSB},$$

with $f_{LSB}^{\Upsilon(nS)}$ the coefficient of the combination, representing the "relative importance" of the LSB distributions with respect to the RSB distributions, with different values for each $\Upsilon(nS)$ state. Intuition demands that $f_{LSB}^{\Upsilon(1S)} > f_{LSB}^{\Upsilon(2S)} > f_{LSB}^{\Upsilon(3S)}$, given that the $\Upsilon(1S)$ peak is the closest to the LSB, while the $\Upsilon(2S)$ and $\Upsilon(3S)$ are further away from the LSB. This intuitive hierarchy of the mixtures corresponds to the hypothesis that the background angular and kinematic distributions change monotonically with dimuon mass. This hypothesis is addressed by studying the angular distributions in detail, as a function of mass.

The polarization extraction framework expects the dimuon events to behave like a vector particle, which can, a priori, not be expected to be the case for continuum dimuon events. Despite this caveat, even though the specific values of $\lambda$ obtained from samples of background events defy any physical interpretation, it can be expected that the functional form (e.g., monotonic, linear behavior) of changes in the polarization

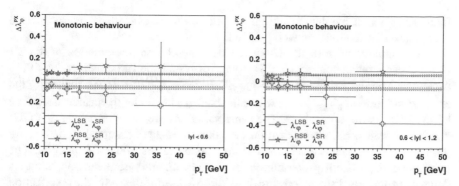

**Fig. 4.11** Differences of $\lambda_\varphi^{LSB}$ (*red*) and $\lambda_\varphi^{RSB}$ (*blue*) with respect to $\lambda_\varphi^{SR}$, in the PX frame for the ranges $|y| < 0.6$ (*left*) and $0.6 < |y| < 1.2$ (*right*) in the non-prompt region. The lines correspond to a fit to the set of points with a constant, and the corresponding uncertainties

parameters, extracted from different background samples, indicates changes of the background angular distributions according to the same functional form. Therefore, an attempt is made to measure the background polarization parameters as a function of dimuon mass, both in a prompt region ($|\ell/\sigma_\ell^{PE}| < 3$), as well as in a non-prompt region ($|\ell/\sigma_\ell^{PE}| > 3$). In the prompt region, this test is limited to the SBs, due to the signal dominance in the mass signal regions. However, given that the $\Upsilon(nS)$ mesons decay promptly, the non-prompt $\Upsilon(nS)$ SR can be considered signal-free, and can be used to determine the background polarization at yet another mass point, allowing for a mass-dependent measurement. These tests suffer from a low number of background events, and are partially statistically inconclusive. However, whenever the statistical precision is good, the polarization parameters measured in the non-prompt SR are in between the results of the parameters in the non-prompt SBs. This is demonstrated in Fig. 4.11 showing the differences of the parameters measured in the PX frame in the non-prompt LSB, $\lambda_\varphi^{LSB}$, and those measured in the non-prompt RSB, $\lambda_\varphi^{RSB}$, with respect to those measured in the non-prompt SR, $\lambda_\varphi^{SR}$, combining the $\Upsilon(nS)$ SRs to increase the statistical power. The values of $\vec{\lambda}$ obtained from the samples in the prompt and non-prompt SB regions are very similar, indicating that the mass-dependent behavior as studied in the non-prompt region can be "extrapolated" to the prompt region. The trends of the results of this study support the hypothesis of monotonically changing background angular distributions as a function of mass.

Therefore, for the nominal analysis a linear relation, the simplest monotonic function, is used to interpolate the LSB and RSB distributions into the SRs. This results in the values $f_{LSB}^{\Upsilon(nS)} = (72, 46, 30\%)$ for the three $\Upsilon(nS)$ states, independent of the kinematic bin. The values of $f_{LSB}^{\Upsilon(nS)}$ are calculated from the medians of the background mass PDF in the LSB, $\Upsilon(nS)$ SRs and RSB.

The study described in Sect. 4.2.3 (bottom two rows of Fig. 4.10), showing that the results for the choices $n_{\sigma_\Upsilon} = 1$ and $n_{\sigma_\Upsilon} = 3$ are compatible, even though the analysis with a mass window of $n_{\sigma_\Upsilon} = 3$ has to cope with considerably larger background fractions, is a valuable demonstration that the assumptions made to interpolate the background distributions are well justified.

The systematic uncertainty associated with the chosen interpolation of the SB distributions is rather conservative. The range of systematic variations for the $\Upsilon(1S)$ background model is chosen to be $[f_{LSB}^{\Upsilon(1S)} - 28\%, f_{LSB}^{\Upsilon(1S)} + 28\%]$, including the extreme variation of $f_{LSB}^{\Upsilon(1S)} = 1$, where the background under the peak is described by the LSB distributions alone. The same variations are calculated for the $\Upsilon(2S)$ and $\Upsilon(3S)$ states. Each result within the allowed range is assumed to be equally probable, and the range is assumed to correspond to the 100% CL interval. The corresponding probability distribution associated with this systematic uncertainty is defined as a uniform distribution with the width $\Delta \vec{\lambda}$, the differences of the polarization parameters obtained from data with the two extreme values of the allowed variations.

An additional component of the systematic uncertainty associated with the background originates from the statistical uncertainty of the background fraction $f_{BG}$ and of the individual entries of the background model histograms. This is taken into account internally in the framework, by drawing a different value of $f_{BG}$ from a Gaussian for each of the $n_{fit}$ repetitions, and by varying the background model histogram bin contents on an event-by-event basis by drawing from an "effective" Poisson distribution, which corresponds to a smearing of the PPD, enlarging its width. The effect of these variations is found to be negligible, by comparing the CL intervals obtained with and without performing these variations during the polarization extraction procedure.

### 4.2.5  Systematic Uncertainties

Each source of systematic uncertainty is evaluated either with toy-MC studies or with studies on data. Systematic variations of $\vec{\lambda}$ are determined, for each of the sources, resulting in an assumed probability distribution describing the effect, in all cases but one a Gaussian distribution, with the systematic variation as the Gaussian width, centered at the most probable value $\hat{\vec{\lambda}}$.

The individual sources of uncertainty are carefully defined to minimize correlations between the individual sources. The definitions of the uncertainties associated with the individual sources have been discussed in the subsections above. Figure 4.12 gives an overview of the kinematic dependence of all individual systematic uncertainties (i.e., their systematic variations corresponding to a 68% CL of the given probability distribution) compared to the respective statistical uncertainty of the measurement, for the $\lambda_{\vartheta}^{HX}$ and $\lambda_{\varphi}^{HX}$ parameters in the $|y| < 0.6$ range, which are very similar to the trends in the $0.6 < |y| < 1.2$ range. The individual contributions in the figures are stacked:

• "Framework Statistics": Framework uncertainties related to the limited number of events (Sect. 4.1.5);

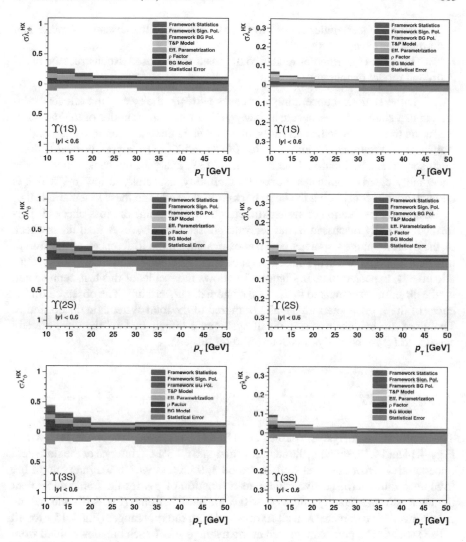

**Fig. 4.12** Summary of all systematic uncertainties considered in the $\Upsilon(1S)$ (*top*), $\Upsilon(2S)$ (*middle*) and $\Upsilon(3S)$ (*bottom*) polarization analyses (*stacked*), compared to the statistical uncertainty (*dark green*), shown for the parameters $\lambda_\vartheta$ (*left*) and $\lambda_\varphi$ (*right*) in the HX frame, in the rapidity range $|y| < 0.6$

- "Framework Sign. Pol.": Framework uncertainties related to large signal polarizations (Sect. 4.1.5);
- "Framework BG Pol.": Framework uncertainties related to the background subtraction procedure (Sect. 4.1.5);
- "$T\&P$ Model": Uncertainties related to the $T\&P$ model assumptions (Sect. 4.2.2);
- "Eff. Parametrization": Uncertainties related to the parametrization of the data $T\&P$ efficiencies, as well as their statistical uncertainties (Sect. 4.2.2);

- "$\rho$ Factor": Uncertainties related to the assumption of the absence of muon-pair correlations (Sect. 4.2.2);
- "BG Model": Uncertainties related to the assumptions used to construct the background model histograms (Sect. 4.2.4);

While the statistical uncertainty increases with $p_T$, the systematic effects evaluated in this study are most severe at low $p_T$. The total uncertainties of the measurements are thus dominated, in all cases, by systematic uncertainties at low $p_T$, and by statistical uncertainties at high $p_T$. In the $\Upsilon(1S)$ and $\Upsilon(2S)$ analyses, no clearly dominant source can be identified, while in the $\Upsilon(3S)$ analysis, the background model uncertainty clearly dominates the total uncertainty, especially at low $p_T$, which is understandable considering the large background fraction in these kinematic cells.

Besides the evaluation of systematic effects, a wide range of cross-checks is performed, partially discussed in the relevant subsections above. A final cross-check of the overall analysis strategy consists of evaluating the differences of the frame-invariant $\tilde{\lambda}$ parameters in the three reference frames, which are, in absence of systematic effects, expected to be 0. Figure 4.13 shows the results of this test, represented by the differences compared to the total systematic uncertainty. The observed differences of the $\tilde{\lambda}$ parameters are small compared to the total systematic uncertainties. Therefore, no evidence of significant systematic effects beyond the ones already accounted for in the analysis is uncovered by this test.

## 4.2.6   Results

The results of this analysis are published in Ref. [1]. A subset of the results is shown in Figs. 4.14 and 4.15, showing the central values and 68.3% CL interval of the statistical uncertainties (error bars), as well as the 68.3, 95.5 and 99.7% CL intervals of the total uncertainties (respective bands), as a function of $p_T$, for the frame-dependent parameters in the rapidity range $|y| < 0.6$ for the HX frame (Fig. 4.14) and for the frame-invariant parameter $\tilde{\lambda}$ in all frames and both rapidity ranges (Fig. 4.15), for all $\Upsilon(nS)$ states. The points are placed at the average $p_T$ of each bin, determined from the raw data distributions (i.e., not corrected for efficiency and acceptance). The full set of results, for all states, parameters and frames can be found in the supplemental material file [25] of Ref. [1], as provided by the journal.

In summary, this analysis is the first measurement of the $\Upsilon(nS)$ polarizations in $pp$ collisions at $\sqrt{s} = 7$ TeV, extending the results with respect to previous measurements at hadron colliders towards higher $p_T$, with higher precision, and with more reliable analysis techniques than most previous analyses of these observables. Both frame-dependent and frame-invariant quantities have been measured,

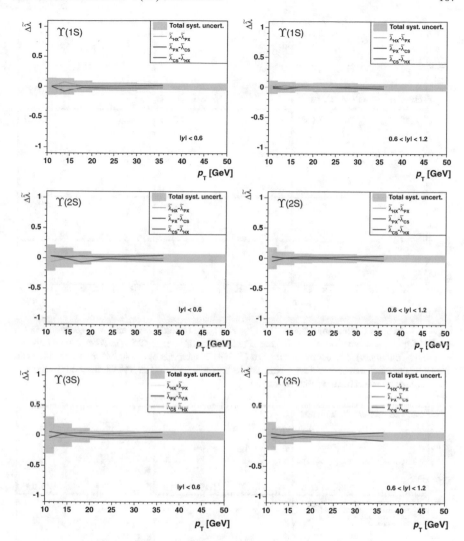

**Fig. 4.13** Differences between the central values of the frame-invariant parameter $\tilde{\lambda}$ of the $\Upsilon(1S)$ (*top*), $\Upsilon(2S)$ (*middle*) and $\Upsilon(3S)$ (*bottom*) analyses in the three reference frames: $\tilde{\lambda}^{CS} - \tilde{\lambda}^{HX}$ (*blue*), $\tilde{\lambda}^{PX} - \tilde{\lambda}^{CS}$ (*red*), $\tilde{\lambda}^{HX} - \tilde{\lambda}^{PX}$ (*green*), in the rapidity ranges $|y| < 0.6$ (*left*) and $0.6 < |y| < 1.2$ (*right*), compared to the total systematic uncertainty (*orange*)

excluding large transverse and longitudinal polarizations in the kinematic regions $|y| < 1.2$ and $10 < p_T < 50\,\text{GeV}$, for all $\Upsilon(nS)$ quarkonia. The results of this analysis play a major role in the interpretation of the LHC quarkonium production cross section and polarization data, discussed in detail in Chap. 5.

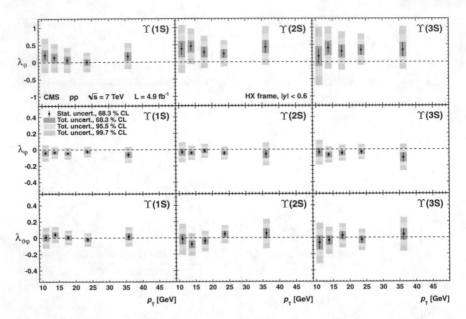

**Fig. 4.14** Results of this analysis for $\lambda_\vartheta$ (*top row*), $\lambda_\varphi$ (*middle row*) and $\lambda_{\vartheta\varphi}$ (*bottom row*) in the HX frame, as a function of $p_T$, for the $\Upsilon(1S)$ (*left column*), $\Upsilon(2S)$ (*middle column*) and the $\Upsilon(3S)$ (*right column*). The central values and 68.3% CL intervals of the statistical uncertainties are shown by the *black markers* and error bars, while the 68.3, 95.5 and 99.7% CL intervals of the total uncertainties are visualized by the colored bands [1]

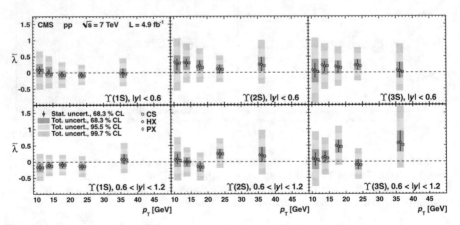

**Fig. 4.15** Results of this analysis for $\tilde\lambda$ in all frames for the rapidity ranges $|y| < 0.6$ (*top row*) and $0.6 < |y| < 1.2$ (*bottom row*), as a function of $p_T$, for the $\Upsilon(1S)$ (*left column*), $\Upsilon(2S)$ (*middle column*) and the $\Upsilon(3S)$ (*right column*) [1]. Central values and uncertainties follow the same definitions as in Fig. 4.14

## 4.3  Measurement of the Prompt $\psi(nS)$ Polarizations

This section describes the measurement of the prompt $\psi(nS)$ polarizations with the CMS detector in $pp$ collisions at $\sqrt{s} = 7\,\text{TeV}$. The corresponding results have been published in Ref. [2]. This analysis profits vastly from the studies and techniques developed for the measurement of the $\Upsilon(nS)$ polarizations, described in Sect. 4.2. Only the main differences with respect to the measurement of the $\Upsilon(nS)$ polarizations are discussed here. More details will be presented in the forthcoming PhD thesis of Ilse Krätschmer, HEPHY, Vienna. The same basic analysis strategy as described in Sect. 4.1 is followed also in this analysis.

### 4.3.1  $\psi(nS)$ Data Processing and Event Selection

The dimuon events for the $J/\psi$ analysis and for the $\psi(2S)$ analysis use different HLT paths, namely *HLT_Dimuon10_Jpsi_Barrel* and *HLT_DimuonX_PsiPrime* (with X=7, 9). The main difference of the $\psi(2S)$ path with respect to the $J/\psi$ and $\Upsilon(nS)$ paths is that the dimuon rapidity was not restricted, allowing for a measurement up to larger values of dimuon $|y|$. The $\psi(nS)$ HLT paths use as input the same L1 trigger seed as is used for the measurement of the $\Upsilon(nS)$ polarizations, *L1_DoubleMu0_HighQ*. The data and particle-gun MC samples for the $\psi(nS)$ analysis are processed in the same way as described in Sect. 4.2.1. The event selection cuts are identical, except for the cut on the lifetime-significance, which is omitted, as the full information about the lifetime distribution is required for the separation of the prompt and non-prompt charmonia (see Sect. 4.3.3).

The polarization is measured in the same rapidity ranges, $|y| < 0.6$ and $0.6 < |y| < 1.2$, adding for the $\psi(2S)$ only a third rapidity range $1.2 < |y| < 1.5$. The $p_\text{T}$ region accessible in this measurement is $14 < p_\text{T} < 70\,\text{GeV}$ for the $J/\psi$ and $14 < p_\text{T} < 50\,\text{GeV}$ for the $\psi(2S)$. This analysis is restricted to a higher-$p_\text{T}$ region ($p_\text{T} > 14\,\text{GeV}$) than the $\Upsilon(nS)$ analysis ($p_\text{T} > 10\,\text{GeV}$), mainly because the decrease of angular acceptance caused by the muon fiducial cuts have a larger effect, for the same dimuon $p_\text{T}$, for lower-mass quarkonia.

### 4.3.2  $\psi(nS)$ Efficiencies

The dimuon efficiency $\epsilon_{\mu\mu}(\vec{p}^{\,\mu^+}, \vec{p}^{\,\mu^-})$ is defined as in Eq. 4.8, using the same parametrized muon efficiencies $\epsilon_\mu^{\text{DataT\&P}}(p_\text{T}^\mu)$ as in the $\Upsilon(nS)$ analysis. Furthermore, the same considerations concerning the vertexing-efficiency $\epsilon_{vtx}$ apply, associated with a negligible systematic uncertainty. The main difference arises from the muon-pair correlations $\rho(\vec{p}^{\,\mu^+}, \vec{p}^{\,\mu^-})$, due to the lower charmonium mass and the higher-$p_\text{T}$ reach of the $J/\psi$ analysis. The $\rho$-factor is studied as a function of $\cos\vartheta^{PX}$ and $\varphi^{PX}$,

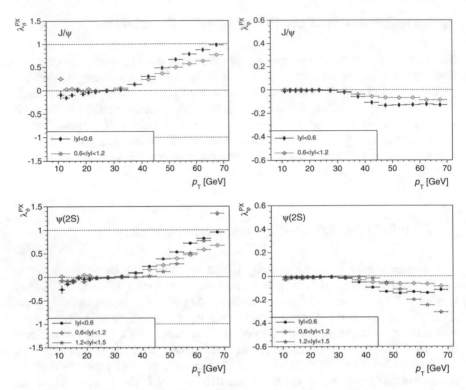

**Fig. 4.16** Effective anisotropy parameters $\lambda_\vartheta^{PX}$ (*left*) and $\lambda_\varphi^{PX}$ (*right*), as a function of $p_T$ and $|y|$, obtained by fitting the J/$\psi$ (*top*) and $\psi(2S)$ (*bottom*) $\rho$-factor angular distributions

in fine bins of dimuon $p_T$,[2] in the rapidity ranges of the analysis, individually for the $\psi(nS)$ states, based on the individual particle-gun MC samples. The "flatness" of the $\rho$-factor angular maps is evaluated as a function of $p_T$ and rapidity, with fits to a PDF based on Eq. 2.5, extracting effective $\vec{\lambda}$ parameters introduced by the muon-pair correlations. Values significantly different from $\vec{0}$ are only observed for the J/$\psi$ analysis for $p_T > 35$ GeV. Despite this observation, the corrections are used for the full kinematic range and both $\psi(nS)$ analyses, to allow for a consistent treatment of the corrections and the systematic uncertainties. Figure 4.16 (top) shows the $p_T$ dependence of the effective anisotropy parameters that are most affected by the muon-pair correlations, $\lambda_\vartheta^{PX}$ and $\lambda_\varphi^{PX}$, for the J/$\psi$ analysis. The results of the corresponding $\psi(2S)$ study are very similar, as shown in the bottom panels of Fig. 4.16, slightly shifting the trend towards higher values of $p_T$ due to the higher $\psi(2S)$ mass. The low-$p_T$ behavior of the results is caused by imperfect modeling of the turn-on curves of the muon efficiencies at very low $p_T$. This observation provides an additional reason to restrict the analysis to the kinematic range $p_T > 14$ GeV.

---

[2] Binning of the $\rho$-factor studies: 20 bins in $\cos \vartheta^{PX}$, 24 bins in $\varphi^{PX}$ and 16 bins in dimuon $p_T$, covering the range $10 < p_T < 70$ GeV.

For the highest $p_T$ values accessible in the $J/\psi$ analysis, the corrections are fairly large, and have to be carefully tested. The $\rho$-factor evaluation from MC can be tested with a data-driven method. A kinematic variable $\Delta R$, depending on $\Delta\eta$, $\Delta\varphi$ and $\Delta p_T$ of the two muons, which shall not be discussed in more detail here, can be defined, which is highly correlated with the distance of the muons in the muon detectors, and thus, the events that are affected by muon-pair correlations can be separated from those where the muons are too far apart to be affected. Tests involving this variable are used to validate the MC $\rho$-factor corrections, further providing a handle on the systematic uncertainty associated with these MC corrections. Both data and MC tests are performed, extracting the polarization parameters from samples in various $\Delta R$-regions, continuously decreasing the muon-pair correlation effects.

### 4.3.3  $\psi(nS)$ Mass and Lifetime Distributions

A major difference of the $\psi(nS)$ analysis is the presence of non-prompt charmonia that originate from decays of heavier B hadrons, mostly from the $B^+$, $B^0$, $B_s$ mesons and the $\Lambda_b$ baryon. This contamination is rather large, even dominating at high $p_T$. The so-called "B fraction" $f_b$, which is the relative contribution of non-prompt $\psi(nS)$ events with respect to the inclusive $\psi(nS)$ events,

$$f_b^{\psi(nS)} = \frac{f_{NP}^{\psi(nS)}}{f_{PR}^{\psi(nS)} + f_{NP}^{\psi(nS)}} \, , \qquad (4.13)$$

increases as a function of $p_T$ and reaches values beyond 65% for the $J/\psi$, with similar trends for the $\psi(2S)$. Already for $p_T = 20\,\text{GeV}$, the B fraction is of the order of 50% [19, 26, 27]. The goal of the analysis is to measure the polarization of the prompt $\psi(nS)$ mesons, the non-prompt contribution being treated as part of the inclusive background and subtracted (see Sect. 4.3.4). Due to the large average decay times of the B hadrons, it is possible to discriminate between prompt and non-prompt charmonia, on a statistical basis, using the lifetime information of the events, as introduced in Sect. 3.2.7.

The three categories of processes contributing to the dimuon data sample, the prompt and non-prompt $\psi(nS)$ charmonia as well as the $\mu\mu$ continuum background, can be separated thanks to both the mass dimension, $M_{\mu\mu}$, as well as the lifetime dimension, $\ell$. The continuum background and the two signal contributions have very different mass shapes, due to the peaking nature of the signals. In the lifetime dimension, one can separate the prompt and non-prompt signal contributions, given that the non-prompt contribution has a large tail towards high values of $\ell$, due to its exponential decay time distribution, while the prompt signal is clustered around $\ell = 0$. The reconstructed $\ell$-values of the prompt signal are different from 0 only because of detector resolution effects, given that the $J/\psi$ and $\psi(2S)$ have average decay times of around $7 \cdot 10^{-21}$ and $2 \cdot 10^{-21}$ s, respectively [21], impossible to be resolved with the CMS tracker.

It is trivial to define a lifetime region at high lifetimes that is dominated by non-prompt signal, with negligible prompt signal contributions. On the other hand, given that in the region around $\ell = 0$ both prompt and non-prompt signals are present, it is not possible to define a lifetime region with negligible non-prompt contributions. However, by defining a lifetime region close to $\ell = 0$, one obtains a region that is dominated by prompt signal, where the fraction of non-prompt charmonia is significantly reduced. This strategy is used in this analysis, which therefore requires the estimation of the relative fractions of the individual contributions in the individual regions.

A two-step fitting approach is employed to estimate the relative fractions of the three contributions in different mass-lifetime regions. First, a fit to the mass dimension is made, then the lifetime dimension is fitted using some inputs from the first step. The details of the fit model will not be discussed in this context.

### $\psi(nS)$ Mass Analysis

The main difference with respect to the $\Upsilon(nS)$ analysis is that the region in between the $\psi(nS)$ states is accessible to study the background distributions, allowing to build background angular distributions from mass sidebands that are much closer to the peak regions, reducing the uncertainty related to the assumptions on the interpolation of these distributions into the peak region. The dimuon mass resolution of the $\psi(nS)$ states ranges from around 20–50 MeV, best at low $p_T$ and mid-rapidity, and worst at high $p_T$ and forward-rapidity, slightly worse for the $\psi(2S)$ with respect to the $J/\psi$, while the $\psi(nS)$ states are separated in mass by 589 MeV [21].

In an initial step, the mass distributions are fit with an unbinned ML method. Figure 4.17 (top) shows examples of the mass distributions and fit results. From these results, three mass regions are defined. The mass region of each state is divided in a $\psi(nS)$ LSB, a $\psi(nS)$ SR and a $\psi(nS)$ RSB, similarly to the definitions in Eq. 4.11, with $n_{\sigma_\psi}^{LSB} = 4.0$, $n_{\sigma_\psi} = 3$ and $n_{\sigma_\psi}^{RSB} = 3.5$. In this analysis, a wider mass region for the signal region is used, characterized by $n_{\sigma_\psi} = 3$. This is affordable, given that the systematic uncertainty associated with the background model is considerably smaller than in the $\Upsilon(nS)$ case, and given that there is no signal cross-feed between the two $\psi(nS)$ states.

### $\psi(nS)$ Lifetime Analysis

In a second step, the lifetime distributions are fit with an unbinned ML method, separately but simultaneously for the three mass regions. The fractions of the continuum background component in the three mass regions are constrained from the mass fit, as calculated by integration of the mass PDFs. Figure 4.17 (bottom) shows examples of the lifetime distributions and fit results, projected in the mass SRs. From the fit results, the average lifetime resolution, $\sigma_\ell$, can be defined, ranging from 10 to 30 μm, best at high $p_T$ and worst at low $p_T$.

The lifetime dimension is divided in two regions, a prompt (PR) region, defined by $\ell \in [-n_{\sigma_\ell} \cdot \sigma_\ell, n_{\sigma_\ell} \cdot \sigma_\ell]$, and a non-prompt (NP) region, defined by $\ell > n_{\sigma_\ell} \cdot \sigma_\ell$,

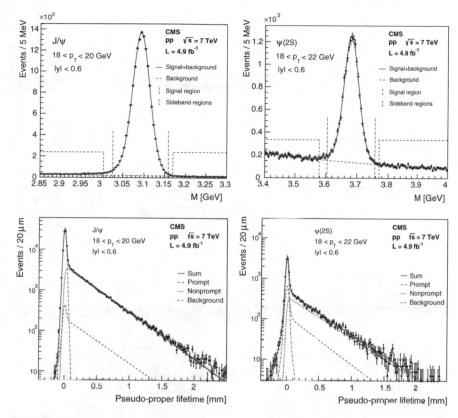

**Fig. 4.17** Dimuon mass distributions (*top*) and lifetime distributions, projected in the mass SR (*bottom*) of the J/$\psi$ (*left*) and $\psi(2S)$ (*right*) analyses, in specific $p_T$ regions in the range $|y| < 0.6$, together with fit results, visualizing the individual contributions, and the definition of the individual mass regions [2]

with $n_{\sigma_\ell} = 3$. In total, the mass-lifetime planes of both $\psi(nS)$ states are divided in 6 regions (PRLSB, PRSR, PRRSB, NPLSB, NPSR and NPRSB). These regions are visualized by the sketch in Fig. 4.18, showing examples of the data distribution and fit results of the mass-lifetime dimensions. At this point, one can calculate by integration of the mass and lifetime PDFs the fractions of all contributions in all mass-lifetime regions.

The reliability of the mass-lifetime fit can be tested by calculating the B fraction as defined in Eq. 4.13, and comparing its $p_T$ dependence to existing results in the literature. The data distributions are very well described in all cases, and the B-fraction results are compatible with previous measurements, validating the procedure. The fractions of the individual components in the prompt J/$\psi$ and $\psi(2S)$ signal regions are shown in Fig. 4.19, for the $|y| < 0.6$ range. The non-prompt contribution can be kept reasonably small with the chosen definition of the PR region. It is below 16% in all cases. The continuum background is rather small for the J/$\psi$, below 6%

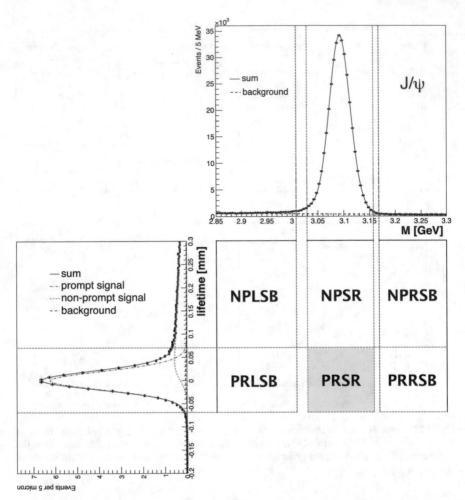

**Fig. 4.18** Sketch of the six individual mass-lifetime regions used in this analysis, illustrated by the $M_{\mu\mu}$ (*top right*) and $\ell$ (*bottom left*) projections of the data and corresponding fit results

in all cases, but reaches values of up to 40% in the worst case for the $\psi(2S)$, in the $1.2 < |y| < 1.5$ range.

Similar to the effects related to cutting on the dimuon mass (see Sect. 4.2.3), a small effect is observed when using a tight lifetime region definition for the PR regions. Studies on MC show that the decay angular distribution is affected by tight cuts on the lifetime variable, distorting the distribution of the opening angle of the two decay muons, which affects the $\cos\vartheta$-$\varphi$ distribution, biasing the extracted polarization parameters. However, the chosen window of the PR lifetime region in the nominal analysis is a $3\sigma_\ell$ window, ensuring that the effect is small. No correction is applied, and the residual differences to results with no cuts, as estimated from MC, are applied as systematic uncertainties.

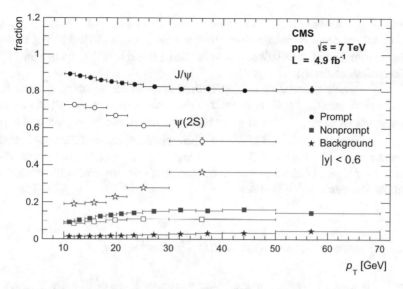

**Fig. 4.19** Relative fractions of the continuum background (*blue*), non-prompt charmonium (*red*) and prompt signal (*black*) contributions in the prompt signal region of the J/$\psi$ (closed markers) and $\psi(2S)$ (open markers), as a function of $p_T$, in the range $|y| < 0.6$ [2]

### 4.3.4 Determination of the $\psi(nS)$ Background Model

Since we are measuring the polarization of the prompt $\psi(nS)$ states, only data from the PRSR is used in the polarization extraction framework. The other regions are used to construct the inclusive background model, defined as a weighted mixture of the angular and kinematic distributions of the non-prompt signal and continuum background components.

The continuum background model is built as a superposition of the distributions in the PRLSB and the PRRSB, as in the $\Upsilon(nS)$ analysis. Again, the background polarizations are studied as a function of mass, and a monotonic trend is observed. Therefore, the same definition is used as in the $\Upsilon(nS)$ analysis, with a $f_{LSB}^{\psi(nS)}$ calculated under the assumption that the background changes linearly as a function of mass. Due to the final state radiation tail of the $\psi(nS)$ mass peaks, the PRLSB contains a non-negligible contamination of signal, between 12 and 22% in the case of the J/$\psi$, and between 2 and 4% in the case of the $\psi(2S)$. The corresponding signal-distributions, approximated by the distributions from the PRSR, are subtracted from the PRLSB distributions. The effect in the PRRSB is negligible (below 5% in the worst case). The effect of this procedure on the estimated values of the polarization parameters is marginal.

The kinematic and angular distributions of the non-prompt charmonium component are modeled from the events in the NPSR, which is dominated by non-prompt signal. However, the remaining background contamination of this sample, below

10% in all J/$\psi$ analysis bins, but up to 40% in the worst case of the $\psi(2S)$ analysis, is taken into account by subtracting, from the distributions in the NPSR region, a linear combination of the distributions in the NPLSB and NPRSB, again using $f_{LSB}^{\psi(nS)}$ as the coefficient for the superposition.

The systematic uncertainty related to the interpolation of the continuum background model into the PRSR is defined in the same way as in the $\Upsilon(nS)$ analysis. The nominal value of $f_{LSB}^{\psi(nS)}$ is close to 50%, and the allowed variation of this parameter is in the range $f_{LSB}^{\psi(nS)} \in [25, 75\%]$. The main quantitative difference is that the distance between the LSB and the RSB is much smaller in the $\psi(nS)$ case, due to the narrower signal peaks, and due to the fact that one can access the background events in between the $\psi(nS)$ states.

### 4.3.5  Summary of the Systematic Uncertainties

Figure 4.20 gives an overview of the kinematic dependence of all individual systematic uncertainties (i.e., their systematic variations corresponding to a 68% CL of the given probability distribution) compared to the respective statistical uncertainty of the measurement, for the $\lambda_\vartheta^{HX}$, $\lambda_\varphi^{HX}$ and $\lambda_{\vartheta\varphi}^{HX}$ parameters in the $|y| < 0.6$ range, which are very similar to the trends in the forward-rapidity ranges.

The systematic uncertainties related to the framework and muon efficiencies are defined in the same way as in the $\Upsilon(nS)$ analysis. The systematic uncertainties related to the $\rho$-factor are not discussed in detail here, some considerations are mentioned in Sect. 4.3.2. The uncertainties related to the definition of the lifetime regions is discussed in Sect. 4.3.3. The individual contributions in the figures are squared and stacked:

- "Lifetime Region": Uncertainties related to the definition of the PR lifetime region (Sect. 4.3.3);
- "Vertexing Eff.": Uncertainties related to the vertexing-efficiency $\epsilon_{vtx}$ (Sects. 4.3.2 and 4.2.2);
- "Eff. Parametrization": Uncertainties related to the parametrization of the data $T\&P$ efficiencies, as well as their statistical uncertainties (Sect. 4.2.2);
- "$T\&P$ Model": Uncertainties related to the $T\&P$ model assumptions (Sect. 4.2.2);
- "Framework": Uncertainties related to the polarization extraction framework (Sect. 4.1.5);
- "$\rho$ Factor": Uncertainties related to the assumptions regarding the muon-pair correlations (Sect. 4.3.2);
- "BG Model": Uncertainties related to the assumptions used to construct the background model histograms (Sect. 4.3.4);

While the statistical uncertainty increases with $p_T$, the systematic effects evaluated in this study are highest at low $p_T$. The total uncertainties of the measurements are dominated by statistical uncertainties at high $p_T$ and by systematic uncertainties at

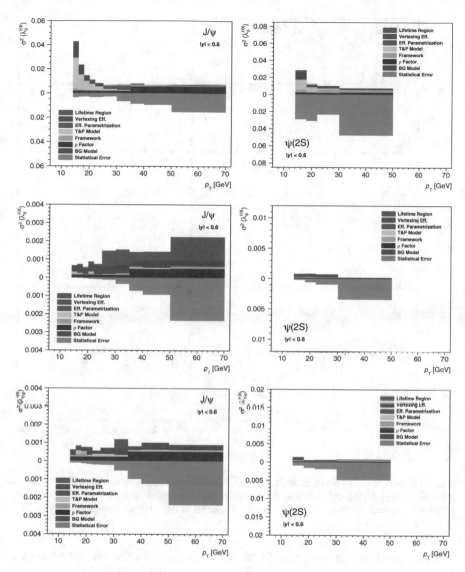

**Fig. 4.20** Summary of all systematic uncertainties considered in the $J/\psi$ (*left*) and $\psi(2S)$ (*right*) polarization analyses (*squared* and *stacked*), compared to the statistical uncertainty (*dark green*), shown for the parameters $\lambda_\vartheta$ (*top*), $\lambda_\varphi$ (*middle*) and $\lambda_{\vartheta\varphi}$ (*bottom*) in the HX frame, in the rapidity range $|y| < 0.6$

low $p_T$, except for a few cases of the $\psi(2S)$, where the total uncertainty is dominated by the statistical uncertainties in all kinematic regions. The dominating sources of uncertainty at low $p_T$ are those related to the muon efficiency, while at high $p_T$, no dominant source can be identified. Figure 4.22 shows that the frame-invariant parameter $\tilde{\lambda}$ is consistent in all the reference frames under study, as in the $\Upsilon(nS)$

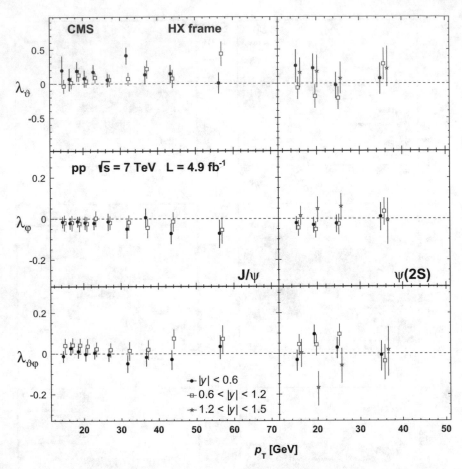

**Fig. 4.21** Results of this analysis for $\lambda_\vartheta$ (*top*), $\lambda_\varphi$ (*middle*) and $\lambda_{\vartheta\varphi}$ (*bottom*) in the HX frame, as a function of $p_T$, for the J/$\psi$ (*left*) and the $\psi(2S)$ (*right*). The central values and 68.3% CL interval of the total uncertainties are shown, for all rapidity ranges of the analysis, in different colors [2]

analysis, providing no evidence of systematic uncertainties that are not accounted for in the analysis.

### 4.3.6  Results

The results of this analysis are published in Ref. [2]. A subset of the results is shown in Figs. 4.21 and 4.22, showing the central values and 68.3% CL interval of the total uncertainties, as a function of $p_T$, for the frame-dependent parameters for all rapidity ranges of the analysis, as measured in the HX frame (Fig. 4.21), and for

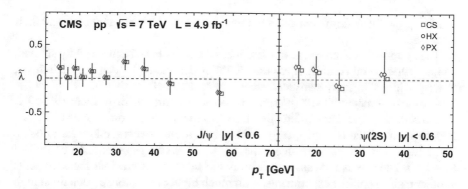

**Fig. 4.22** Results of this analysis for $\tilde{\lambda}$ in all frames for the rapidity range $|y| < 0.6$, as a function of $p_{\mathrm{T}}$, for the $J/\psi$ (*left*) and the $\psi(2S)$ (*right*). The central values and 68.3% CL interval of the total uncertainties are shown, as evaluated in the reference frames of the analysis, in different colors [2]

the frame-invariant parameter $\tilde{\lambda}$ in all frames at mid-rapidity (Fig. 4.22), for both $\psi(nS)$ states. The points are placed at the average $p_{\mathrm{T}}$ of each bin, determined from the raw data distributions (i.e., not corrected for efficiency and acceptance). The full set of results, for all states, parameters and frames can be found in the supplemental material file [28] of Ref. [2], as provided by the journal.

In summary, this analysis measures the polarizations of the prompt $\psi(nS)$ states in $pp$ collisions at $\sqrt{s} = 7$ TeV, extending the results of previous measurements at hadron colliders towards higher $p_{\mathrm{T}}$, with higher precision, and with more reliable analysis techniques than the previous analyses of these observables. Both frame-dependent and frame-invariant quantities have been measured, excluding large transverse and longitudinal polarizations in the kinematic regions $|y| < 1.2$ and $14 < p_{\mathrm{T}} < 70$ GeV for the $J/\psi$, and in the kinematic regions $|y| < 1.5$ and $14 < p_{\mathrm{T}} < 50$ GeV for the $\psi(2S)$. The results of this analysis play a major role in the interpretation of the LHC quarkonium production cross section and polarization data, discussed in detail in Chap. 5.

## 4.4   Data Analysis Summary

The polarizations of all five S-wave quarkonium states have been measured, based on CMS data collected in $pp$ collisions at 7 TeV. The analyses have strictly followed the suggestions of improved analysis techniques for quarkonium polarization measurements, as introduced in Sect. 2.3, exploiting the full information of the 2-dimensional decay angular distributions, measuring the frame dependent parameters $\lambda_{\vartheta}$, $\lambda_{\varphi}$ and $\lambda_{\vartheta\varphi}$, in the Helicity, Collins-Soper and Perpendicular-Helicity frames, in addition to the measurement of the frame-invariant parameter $\tilde{\lambda}$.

Novel and well performing Markov Chain Monte Carlo techniques have been developed and validated, for the background subtraction as well as for the determination of the full posterior probability density function of the anisotropy parameters, providing the full information about the measured polarization parameters and their correlations.

The systematic uncertainties of these measurements have been carefully studied, and are under good control. These uncertainties dominate the total uncertainties at low $p_T$, while at high $p_T$ statistical uncertainties dominate.

The measurements of the $\Upsilon(nS)$ polarizations cover the kinematic range $|y| < 1.2$, in two cells in $|y|$, and are differential in $p_T$, up to a $p_T$ of the order of 35 GeV. The measurement of the prompt $J/\psi$ ($\psi(2S)$) polarizations covers the kinematic range $|y| < 1.2$ ($|y| < 1.5$), in two (three) cells in $|y|$, and reach a $p_T$ of the order of 55 GeV (35 GeV). The results increase significantly the $p_T$ reach of previous measurements at other hadron collider experiments, with much higher precision, especially at high $p_T$.

None of the S-wave quarkonium states show significant polarizations, neither transverse nor longitudinal, providing vital information for the interpretation of LHC quarkonium production results, as discussed in Chap. 5.

# References

1. CMS Collaboration (2013) Measurement of the $\Upsilon(1S)$, $\Upsilon(2S)$ and $\Upsilon(3S)$ polarizations in $pp$ collisions at $\sqrt{s} = 7$ TeV. Phys Rev Lett 110:081802
2. CMS Collaboration (2013) Measurement of the prompt $J/\psi$ and $\psi(2S)$ polarizations in pp collisions at $\sqrt{s} = 7$ TeV. Phys Lett B 727:381
3. Faccioli P, Seixas J, Knünz V (2012) A new procedure for the determination of angular distribution parameters in dilepton vector meson decays. CMS AN-11-535 (Internal Note, Unpublished)
4. Wöhri H, Krätschmer I, Zhang L, Wang J (2012) T&P single muon efficiencies for low $p_T$ dimuon triggers in 2011. CMS AN-11-417 (Internal Note, Unpublished)
5. Wöhri H, Krätschmer I, Zhang L, Knünz V et al (2012) Low $p_T$ muon and dimuon efficiencies. CMS AN-12-088 (Internal Note, Unpublished)
6. Knünz V, Faccioli P, Lourenço C, Krätschmer I et al (2012) Measurement of the $\Upsilon(1S)$, $\Upsilon(2S)$ and $\Upsilon(3S)$ polarizations in $pp$ collisions at $\sqrt{s} = 7$ TeV. CMS AN-12-140 (Internal Note, Unpublished)
7. Krätschmer I, Zhang L, Knünz V, Wöhri H et al (2013) $J/\psi$ and $\psi(2S)$ polarizations in $pp$ collisions at $\sqrt{s} = 7$ TeV. CMS AN-13-016 (Internal Note, Unpublished)
8. Faccioli P, Lourenço C, Seixas J, Wöhri H (2009) J/$\psi$ polarization from fixed-target to collider energies. Phys Rev Lett 102:151802
9. Faccioli P, Lourenço C, Seixas J (2010) Rotation-invariant relations in vector meson decays into fermion pairs. Phys Rev Lett 105:061601
10. Faccioli P, Lourenço C, Seixas J (2010) A new approach to quarkonium polarization studies. Phys Rev D 81:111502
11. Faccioli P, Lourenço C, Seixas J, Wöhri H (2011) Model-independent constraints on the shape parameters of dilepton angular distributions. Phys Rev D 83:056008
12. Faccioli P, Lourenço C, Seixas J, Wöhri H (2010) Towards the experimental clarification of quarkonium polarization. Eur Phys J C 69:657
13. Faccioli P, Lourenço C, Seixas J, Wöhri H (2011) Determination of $\chi_c$ and $\chi_b$ polarizations from dilepton angular distributions in radiative decays. Phys Rev D 83:096001
14. James F, Winkler M (2004) MINUIT user's guide. CERN, Geneva. http://seal.web.cern.ch/seal/documents/minuit/mnusersguide.pdf

15. Knünz V, Adam W, Frühwirth R, Strauss J (2011) Studies of the $J/\psi$ polarization fit with a toy Monte Carlo program. CMS AN-11-087 (Internal Note, Unpublished)
16. CMS Collaboration (2011) Measurements of inclusive $W$ and $Z$ cross sections in pp collisions at $\sqrt{s} = 7$ TeV. J High Energy Phys 1101:080
17. Hastings WK (1970) Monte Carlo sampling methods using Markov chains and their applications. Biometrika 57:97
18. HERA-B Collaboration (2009) Kinematic distributions and nuclear effects of $J/\psi$ production in 920-GeV fixed-target proton-nucleus collisions. Eur Phys J C 60:525
19. CMS Collaboration (2012) $J/\psi$ and $\psi(2S)$ production in $pp$ collisions at $\sqrt{s} = 7$ TeV. J High Energy Phys 1202:011
20. CMS Collaboration (2013) Measurement of the $\Upsilon(1S)$, $\Upsilon(2S)$, and $\Upsilon(3S)$ cross sections in pp collisions at $\sqrt{s} = 7$ TeV. Phys Lett B 727:101
21. Olive KA et al (2014) Particle data group. Chin Phys C 38:090001
22. Sjostrand T, Mrenna S, Skands PZ (2006) PYTHIA 6.4 physics and manual. J High Energy Phys 0605:026
23. Krätschmer I (2012) Understanding Muon Detection Efficiencies for Quarkonium Polarization Measurements at the Compact Muon Solenoid. Diploma thesis, Institute of High Energy Physics, Vienna University of Technology
24. Drell SD, Yan TM (1970) Massive lepton pair production in hadron–hadron collisions at high-energies. Phys Rev Lett 25:316
25. Supplemental Material. Rev Lett 110:081802 (2013)
26. ATLAS Collaboration (2011) Measurement of the differential cross-sections of inclusive, prompt and non-prompt $J/\psi$ production in proton–proton collisions at $\sqrt{s} = 7$ TeV. Nucl Phys B 850:387
27. ATLAS Collaboration (2014) Measurement of the production cross-section of $\psi(2S) \rightarrow J/\psi(\rightarrow \mu^+\mu^-)\pi^+\pi^-$ in pp collisions at $\sqrt{s} = 7$ TeV at ATLAS. J High Energy Phys 1409:79
28. Supplemental Material of Phys Lett B 727:381 (2013)

# Chapter 5
# Discussion of Results

The LHC experiments have provided essential inputs to solve the problems in the understanding of quarkonium production at a fundamental level. This chapter aims at summarizing the progress made in the LHC era so far, both from the experimental and theoretical perspectives. The wealth of available quarkonium production data from the LHC experiments is summarized in Sect. 5.1, followed by a summary of the main NRQCD analyses, comparing the data to the state-of-the-art theory calculations in Sect. 5.2, which displays the still unsatisfactory situation in the understanding of quarkonium production. Furthermore, a data-driven approach [1] to reconcile the available LHC data with NRQCD calculations is described in detail in Sect. 5.3, providing a straightforward solution to the quarkonium polarization puzzle.

## 5.1 Quarkonium Production Data at the LHC

Besides the measurements discussed in the previous chapter, a wealth of new quarkonium production results from the LHC experiments is available, as well as Tevatron measurements, using more sophisticated and reliable analysis methodologies. This section lists the relevant measurements, providing some comparison figures of the CMS measurements with those of the other experiments. From this information, several data-driven observations can be made, which are discussed here.

### 5.1.1 Cross Section Measurements

The results discussed here are an incomplete list of LHC quarkonium cross section measurements, focussing on the results that are most relevant for this thesis, restricted to measurements in $pp$ collisions at $\sqrt{s} = 7$ and 8 TeV.

© Springer International Publishing Switzerland 2017
V. Knünz, *Measurement of Quarkonium Polarization to Probe QCD at the LHC*, Springer Theses, DOI 10.1007/978-3-319-49935-2_5

In the S-wave quarkonium production sector, CMS has published measurements of the prompt $J/\psi$ and $\psi(2S)$ production cross sections, based on a fraction of the 7 TeV data collected in 2010 [2], and on the full 2010 [3] and 2011 [4, 5] data sets, extending the $p_T$ reach of the results up to above $100\,\mathrm{GeV}^1$ for the $J/\psi$ and to above $80\,\mathrm{GeV}$ for the $\psi(2S)$. Moreover, CMS has published measurements of the $\Upsilon(nS)$ production cross sections, based on a fraction of the 7 TeV data collected in 2010 [6], and on the full 2010 [7] and 2011 [8, 9] data sets, extending the $p_T$ reach up to above $80\,\mathrm{GeV}$. ATLAS has measured the $J/\psi$ and $\psi(2S)$ production cross sections at 7 TeV [10, 11], reaching maximum $p_T$ ranges of around $40\,\mathrm{GeV}$ and around $70\,\mathrm{GeV}$, respectively. The results of the $\Upsilon(nS)$ cross sections at 7 TeV [12] have a $p_T$ reach of around $60$–$65\,\mathrm{GeV}$. LHCb has measured the prompt production cross sections of the $\psi(nS)$ and $\Upsilon(nS)$ states at 7 TeV [13–15], as well as, partially, at 8 TeV [16]. These measurements are restricted to the rapidity region $2 < y < 4.5$, and to $p_T < 15\,\mathrm{GeV}$. ALICE has measured the inclusive [17] and prompt [18] cross sections of the $J/\psi$ at 7 TeV, covering the rapidity ranges $|y| < 0.9$ and $2.5 < y < 4$, and $p_T < 8\,\mathrm{GeV}$.

In the P-wave quarkonium production sector, CMS was so far limited to the measurement of the relative prompt $\chi_{c2}/\chi_{c1}$ [19] and $\chi_{b2}(1P)/\chi_{b1}(1P)$ [20, 21] production cross section ratios, at 7 and 8 TeV, respectively, due to the difficulty of estimating the photon conversion reconstruction efficiency, which largely cancels in the cross section ratio measurements. The $\chi_{b2}(1P)/\chi_{b1}(1P)$ analysis was the first of its kind at a hadron collider, profiting from the extremely good $M^\chi$ mass resolution at CMS. Also ATLAS has a successful P-wave quarkonium physics program. They were the first among the LHC experiments to see significant $\chi_b(3P) \rightarrow \Upsilon(1, 2S)$ decays [22]. Moreover, ATLAS published an extensive production analysis of the $\chi_c$ system at 7 TeV [23], including the measurement of the prompt $\chi_{c2}/\chi_{c1}$ cross section ratio, the $\chi_{c1,2} \rightarrow J/\psi + \gamma$ feed-down fractions, the $\chi_{c1,2}$ B fractions, as well as the production cross sections of the prompt $\chi_{c1}$ and $\chi_{c2}$ states. LHCb also conducted a series of measurements of the P-wave states, including the measurement of the prompt $\chi_{c2}/\chi_{c1}$ production cross section ratio at 7 TeV [24, 25], the measurement of the $\chi_{b2}(1P)/\chi_{b1}(1P)$ production cross section ratio at 8 TeV [26], as well as feed-down fraction measurements of the $\chi_c \rightarrow J/\psi + \gamma$ at 7 TeV [27], $\chi_b(1P) \rightarrow \Upsilon(1S) + \gamma$ at 7 TeV [28, 29] and 8 TeV [29], and of the $\chi_b(2P) \rightarrow \Upsilon(1, 2S) + \gamma$ and $\chi_b(3P) \rightarrow \Upsilon(nS) + \gamma$ ($n = 1, 2, 3$) feed-down decays at 7 and at 8 TeV [29].

Figure 5.1 shows a comparison of the $p_T$ differential production cross sections of a subset of the measurements at 7 TeV as discussed above, restricted to measurements at mid-rapidity, to facilitate the comparison. Cross sections of the S-wave $\psi(nS)$, $\Upsilon(nS)$ and the P-wave $\chi_{cJ}$ states are shown. This compilation of measurements therefore includes states that are characterized by very different masses. In order to be able to compare the shapes of the cross sections, a mass-rescaling is applied on the $p_T$ variable, showing the results as a function of $p_T/M_\varrho$, to compensate the effects of different average parton momenta and phase spaces [1], where $M_\varrho$ is the mass of the measured quarkonium state.

---

[1]The $p_T$ reach is given by the average $p_T$ of the highest $p_T$ bin of a measurement.

**Fig. 5.1** Comparison of the $p_T$ differential production cross sections of a subset of the measurements as discussed in the text, restricted to measurements at mid-rapidity, as a function of $p_T/M_Q$. Additionally, a fitted empirical power-law function is shown, with different normalization for each quarkonium state [1]

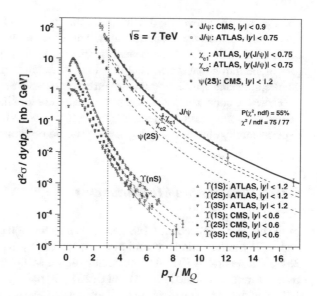

It can be seen that above a certain value of $p_T/M_Q$, the shapes of the cross sections are very similar. This finding is quantified in the following way. The measurement characterized by the smallest uncertainties and the highest $p_T/M_Q$ reach, the CMS $J/\psi$ measurement [3], is fit with an empirical power-law function [30],

$$\frac{d^2\sigma}{dp_T dy} \propto \frac{p_T}{M_Q}\left[1 + \frac{1}{\beta-2}\cdot\frac{(p_T/M_Q)^2}{\gamma}\right]^{-\beta},\tag{5.1}$$

in the region $p_T/M_Q > 3$. The fitted shape parameters are then fixed, and the normalizations are fit to each of the other curves displayed in the figure. The resulting description of all cross section data points with $p_T/M_Q > 3$ is very good, with a $\chi^2$-probability of 55%. This study shows that the $p_T$-differential quarkonium cross sections, as a function of $p_T/M_Q$, for sufficiently large $p_T/M_Q$ values, for all $\psi(nS)$, $\Upsilon(nS)$ and $\chi_{c1,2}$ states can be described by one simple empirical function, with the same shape for each quarkonium state, within the uncertainties of the current measurements.

The individual states of this study include S-wave and P-wave states, with very different feed-down characteristics, as well as charmonium and bottomonium states. This study suggests that quarkonium production should, in fact, be rather simple, and proceed dominantly via one single production mechanism, identical for all quarkonium states. This is the first of two main data-driven observations discussed in this chapter. Given that S-wave and P-wave states have very different color-singlet channels, $^1S_0^{[1]}$ and $^3P_J^{[1]}$, respectively, this observation indicates that the CS contributions are negligible, for both S-wave and P-wave production. The dominating production mechanism should therefore be one single color-octet transition. The data-driven observation still allows for a democratic mixture of CO processes if the mixture is very similar for all quarkonium states. This is rather unlikely, considering that the different masses of the component quarks, as well as the different binding energies of the individual states, should lead to different non-perturbative effects for all quarkonium states, affecting the mixture of processes.

These are conclusions that can be drawn before looking at any quarkonium polarization data, and before conducting any complicated NRQCD fitting analyses. This experimentally observed $p_T/M_Q$ scaling should be confirmed with higher precision.

### 5.1.2  Polarization Measurements

In the charmonium sector, CMS has measured the polarizations of the prompt $J/\psi$ and $\psi(2S)$ polarizations at mid-rapidity, with a $p_T$ reach of the order of 60 and 35 GeV, respectively [31]. The kinematic region accessed by CMS is complemented by measurements of the $J/\psi$ [32] and $\psi(2S)$ [33] polarizations at forward-rapidity by LHCb, with a $p_T$ reach of around 15 GeV, and a measurement of the $J/\psi$ polarization at forward-rapidity and very low $p_T$ by ALICE [34]. A subset of the results of these analyses is summarized in Fig. 5.2, showing the results of the frame-dependent polarization parameters $\lambda_\vartheta$, $\lambda_\varphi$ and $\lambda_{\vartheta\varphi}$, as measured in the HX frame, as well as the frame-invariant parameter $\tilde{\lambda}$, as a function of $p_T$, for all rapidity ranges considered in the analyses.

In the bottomonium sector, CMS has measured the polarizations of the $\Upsilon(nS)$ states at mid-rapidity [35], with a $p_T$ reach of around 35 GeV, complementing an updated measurement of the $\Upsilon(nS)$ polarizations from CDF [36], also at mid-rapidity but covering lower $p_T$ regions, reaching around 25 GeV. The CMS measurement is affected by considerably smaller total uncertainties with respect to the CDF measurement, especially at high $p_T$. The results of CMS are compared to the results of the corresponding CDF analysis in Fig. 5.3, showing the results of the frame-dependent polarization parameters $\lambda_\vartheta$, $\lambda_\varphi$ and $\lambda_{\vartheta\varphi}$, as measured in the HX frame, as well as the frame-invariant parameter $\tilde{\lambda}$, as a function of $p_T$, for all rapidity ranges considered in the analyses.

All measurements of quarkonium polarization discussed here, with the exception of the $J/\psi$ measurement from ALICE [34], employ the multi-dimensional and frame-invariant analysis methodology that is discussed in Sect. 2.3, and refer to prompt quarkonia. All measurements are based on data collected in $pp$ collisions at $\sqrt{s} = 7$ TeV, except for the $\Upsilon(nS)$ polarization measurement from CDF, based on $p\bar{p}$ collisions at $\sqrt{s} = 1.96$ TeV. The comparison figures of the polarizations of the individual measurements are not entirely fair, given the different CM energy in the $\Upsilon(nS)$ case, and the various rapidity regions of the measurements in the $\psi(nS)$ case. However, from NRQCD calculations [37], one knows that no large rapidity or $\sqrt{s}$-dependencies are expected to be observed for the shape of the angular distributions. On the other hand, the different kinematics of the individual measurements

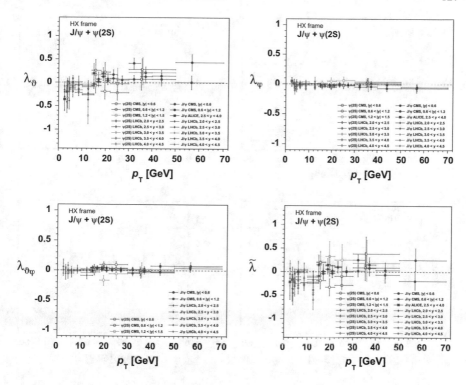

**Fig. 5.2** Comparison of the results of this thesis [31] (*round markers*), with the results of the corresponding LHCb [32, 33] (*diamond markers*) and ALICE [34] (*square markers*) analyses, in the HX frame as a function of $p_T$, for the parameters $\lambda_\vartheta$ (*top left*), $\lambda_\varphi$ (*top right*), $\lambda_{\vartheta\varphi}$ (*bottom left*) and $\tilde{\lambda}$ (*bottom right*) for the $J/\psi$ (*closed markers*) and the $\psi(2S)$ (*open markers*). The uncertainties correspond to the total uncertainties at 68.3% CL. The results of the individual rapidity ranges are distinguished by various colors

can affect the interpretation of the frame-dependent polarization parameters, as the definitions of the reference axes depend on $y$ (see Sect. 2.3.1). Therefore, the comparison which is least affected by the differences of the individual measurements is the one using the frame-invariant parameter $\tilde{\lambda}$.

Contemplating the overall situation of the experimental information about the quarkonium polarization observables in the LHC era, including the updated Tevatron results, there are two data-driven observations to be discussed. After the experimental ambiguities and inconsistencies in the pre-LHC era in the area of quarkonium polarization, it is very satisfying to observe that the LHC has provided a clear experimental picture of quarkonium polarization, with consistent measurements throughout the individual experiments and states. This change of the quality of the quarkonium polarization data is on one hand a consequence of the improvements of the understanding of quarkonium polarization and the resulting recipes for reliable measurements of quarkonium polarization (as discussed in Sect. 2.3), and on the other hand, it is cer-

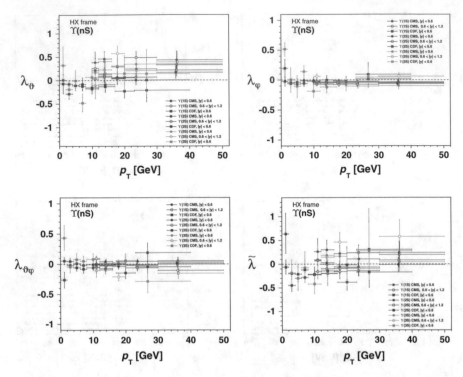

**Fig. 5.3** Comparison of the results of this thesis [35] (*round markers*), with the results of the corresponding CDF analysis [36] (*square markers*), in the HX frame as a function of $p_T$, for the parameters $\lambda_\vartheta$ (*top left*), $\lambda_\varphi$ (*top right*), $\lambda_{\vartheta\varphi}$ (*bottom left*) and $\tilde{\lambda}$ (*bottom right*) for the $\Upsilon(1S)$ (*blue*), $\Upsilon(2S)$ (*red*) and the $\Upsilon(3S)$ (*green*). The uncertainties correspond to the total uncertainties at 68.3% CL. The results of the $|y| < 0.6$ ($0.6 < |y| < 1.2$) range are shown with *closed* (*open*) markers

tainly also caused by the very careful study of systematic effects by the individual experiments conducting these highly non-trivial measurements.

Certainly, one can spot small discrepancies, but the overall picture is very consistent. For example, the $\lambda_\vartheta^{HX}$ results for the $\Upsilon(1S)$ state of the CDF analysis are systematically below the corresponding CMS results, albeit mostly covered by the large uncertainties of the CDF points. Regarding the $\tilde{\lambda}$ parameter results, carefully comparing the values in the Collins-Soper and HX frames, as measured by CDF [36], differences can be spotted that might hint at residual systematic effects not accounted for in the analysis. Another such example is the highest-$p_T$ point in the most forward-rapidity LHCb $\psi(2S)$ result of $\lambda_\vartheta^{HX}$, which is around $\lambda_\vartheta = -0.7 \pm 0.2$, a rather large value albeit less than three standard-deviations away from the general trend observed in this region. The corresponding effect in $\tilde{\lambda}$ is even smaller. This result should be interpreted as a fluctuation.

The LHC experiments, with a fairly large share of the CMS experiment, have provided a clear experimental pattern of quarkonium polarization in hadron collisions. No large polarizations are observed in any of the S-wave quarkonium states, neither longitudinal, nor transverse, in none of the considered frames, the results clustering around the unpolarized limit. This is true for all S-wave quarkonium states, which are affected by very different fractions of feed-down decays of heavier quarkonium states. No significant dependencies on the quarkonium $p_T$ and rapidity are observed. No significant differences between charmonium and bottomonium states are seen. These observations support the data-driven observation formulated in Sect. 5.1.1, strengthening the conjecture that all quarkonium states are produced in a very simple and similar way, either by one dominating CO mechanism, or by mixtures of various CO mechanisms, very similar for all quarkonium states.

Contrary to the shape of the $p_T$-differential cross sections, the polarization observables allow for an immediate interpretation concerning the preferred angular momentum eigenstates. Given the fact that all measured polarizations are very close to the unpolarized case, the straightforward explanation is that the CO channel through which all quarkonia are dominantly produced has to be the $^1S_0^{[8]}$ state.

## 5.2 NRQCD Analyses Review

The data of the LHC era, together with the data of the pre-LHC era, can be used to update the estimations of the LDMEs, using the theoretical inputs introduced in Sect. 2.2.1. The situation has changed, mostly due to the availability of high-quality quarkonium polarization measurements, but also due to quarkonium production cross section measurements extending to much higher values of $p_T$ than those used in the pre-LHC era to fit the LDMEs. There are several groups performing such LDME-fits, also denoted here as "NRQCD analyses", which are briefly introduced in this section, in the interest of putting in order the possibly confusing situation in the literature. There are three main groups performing these NRQCD analyses. The German group associated with the calculations and analyses described in Refs. [37–39] is denoted here as "BK",[2] while the Chinese groups associated with Refs. [40–43] are denoted here as "GWWZ" ("CMSWZ").

### BK NRQCD Analyses

The BK group provides $J/\psi$ hadroproduction SDC calculations at NLO for the CS and the individual CO channels [38], complemented by the corresponding calculations for the polarization observables [37]. With these theory inputs, a NRQCD analysis is performed [39], considering $J/\psi$ cross section measurements from various experiments and collision systems, a so-called "global fit", including hadroproduction data from the RHIC, Tevatron and LHC experiments, as well as photoproduction data measured at the HERA experiments, and $e^+e^-$ data from LEP and KEKB. The goal of this analysis is to estimate the $J/\psi$ LDMEs of the $^1S_0^{[8]}$, $^3S_1^{[8]}$ and $^3P_J^{[8]}$ CO channels, by fitting a superposition of SDCs to the chosen $p_T$ differential cross section data, according to the NRQCD factorization formalism introduced in Eq. 2.1. The

---

[2]The chosen notation reflects the initials of the surnames of the authors.

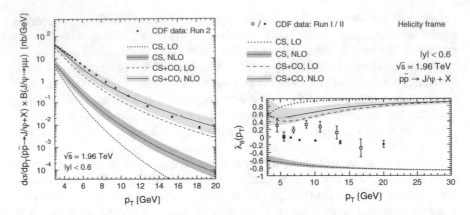

**Fig. 5.4** NRQCD calculations for the direct $J/\psi$ $p_T$-differential production cross section [39] compared to prompt CDF data [44] (*left*). NRQCD calculations for the direct $J/\psi$ polarization parameter $\lambda_\vartheta$ of directly produced $J/\psi$ [37], in the HX frame, as a function of $p_T$, compared to prompt CDF data [46, 47] (*right*). *Yellow and cyan bands* represent the theory uncertainties of the CS and the inclusive CS+CO channels, respectively

selected data requires $p_T > 3$ GeV for hadroproduction data and $p_T > 1$ GeV for photoproduction and $e^+e^-$ data.

The $J/\psi$ cross section data of several collision systems can be reproduced reasonably well by this NRQCD analysis, most data points being compatible with the calculations, within the large theoretical uncertainties. However, theoretical uncertainties are not considered in the fit, and only calculated a posteriori.

The authors claim that this analysis "consolidates the verification of NRQCD factorization for charmonium and provides rigorous evidence for LDME universality and the existence of CO processes in nature" [39]. Figure 5.4 (left) shows the comparison of the CDF measurement [44] to the fit result. While the data are within the theoretical uncertainties (yellow band), the shapes of the data and fit result are clearly very different. Given that the theoretical uncertainties mostly reflect effects that affect the normalization and not the shape of the SDCs, the claimed consistency of their fit result and the data should be viewed with a grain of salt.

In a following paper [37], BK calculate the polarization observables in the HX frame, using as input the LDMEs as estimated in Ref. [39], and compare the polar anisotropy to the CDF data, as shown in Fig. 5.4 (right). The transversely polarized $^3S_1^{[8]}$ and $^3P_J^{[8]}$ channels are the dominating contributions, especially at high $p_T$, resulting in a $J/\psi$ polarization that is almost fully transverse, increasing with $p_T$, inconsistent with the two CDF measurements. It has to be emphasized that no feed-down contributions are taken into account in this analysis and in the calculation of the polarization observables, while the comparison can only be done, currently, with prompt $J/\psi$ data.

With the same strategy, BK performs fits estimating $\psi(2S)$ LDMEs [45]. The results of this study are very similar to the $J/\psi$ analysis, with the $^3S_1^{[8]}$ channel completely dominating, leading to almost fully transverse polarization at high $p_T$.

**CMSWZ NRQCD Analyses**

The CMSWZ group provides an independent calculation of the SDCs [42] and polarization observables [43] of the CS and the individual CO states. An independent extraction of the $J/\psi$ LDMEs is attempted with a fit to hadroproduction data only, restricted to CDF measurements. Contrary to BK, the CMSWZ group includes the CDF Run II $J/\psi$ polarization measurement [47]. Data points with $p_T > 7\,\text{GeV}$ are selected, due to "existing non-perturbative effects" [43] in the low-$p_T$ region.

No feed-down contributions are taken into account, the fit therefore compares prompt $J/\psi$ data with direct $J/\psi$ calculations, as is the case in the BK analyses, which can especially affect the polarization observables [40, 48]. The fit yields a surprising result. The strongly transverse $^3P_J^{[8]}$ channel is found to contribute to the color-inclusive cross section with a large negative partial cross section, effectively corresponding to a strongly longitudinal contribution. The transverse $^3S_1^{[8]}$ contribution is positive and large, and almost fully cancels the longitudinal negative $^3P_J$ component, resulting in an unpolarized or even slightly longitudinally polarized directly produced $J/\psi$. With this result, the small polarizations measured at CDF can be reproduced.

This group has recently performed an extensive NRQCD analysis in the bottomonium system [49], taking into account all feed-down contributions, including both cross section and polarization data with the requirement of $p_T > 15\,\text{GeV}$ in their fitting procedure. Due to its recent appearance, this analysis is not further considered in this thesis.

**GWWZ NRQCD Analyses**

The GWWZ group provides NLO NRQCD calculations of quarkonium cross sections and polarizations for hadroproduction through the CS and the individual CO channels, for the charmonium [40] and $\Upsilon(nS)$ [41] states. These calculations (for the $J/\psi$ only) could be successfully compared to the corresponding calculations of BK [37] and CMSWZ [40], reproducing their fit results when using their strategies. This group also performs NRQCD fits, extracting several sets of LDMEs of the charmonium and bottomonium states. These are the first NLO NRQCD analyses that take into account the feed-down contributions of the heavier quarkonium states. This is an important step in the clarification of quarkonium production.

The charmonium analysis [40] is based on hadroproduction data from CDF and LHCb, imposing a minimum $p_T$ requirement, $p_T > 7\,\text{GeV}$, as "It is known that the double expansion in $\alpha_s$ and $v^2$ is not good enough in the small $p_T$ regions" [41]. No polarization data are used, and no theoretical uncertainties are considered in the fits. The analysis is organized as an iterative procedure. First, the $\psi(2S)$ LDMEs for the $^1S_0^{[8]}$, $^3S_1^{[8]}$ and $^3P_J^{[8]}$ channels are estimated by fits to CDF and LHCb production cross section measurements. Then, the $\chi_c$ $^3S_1^{[8]}$ LDME is estimated by fitting the $\chi_c \to J/\psi + \gamma$ feed-down fractions measured at CDF and LHCb, not taking into account the radiative feed-down decays $\psi(2S) \to \chi_{cJ} + \gamma$. Finally, the LDMEs of the $J/\psi$ are obtained by fitting CDF and LHCb $J/\psi$ production cross section measurements, taking into account the feed-down contributions as fixed in the previous

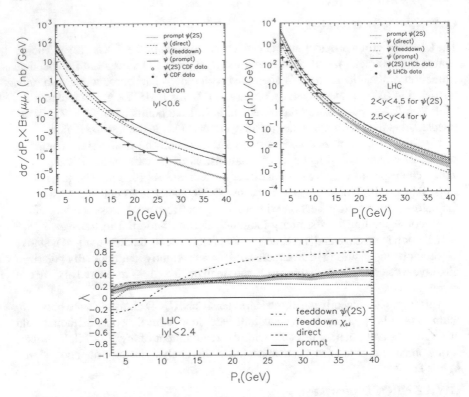

**Fig. 5.5** NRQCD calculations for prompt $J/\psi$ and $\psi(2S)$ $p_T$-differential production cross sections [40], compared to prompt CDF (*top left*) and LHCb data (*top right*), and calculations of $\lambda_\vartheta^{HX}$ for the prompt $J/\psi$ (*bottom*)

two steps of the analysis. Figure 5.5 (top) shows fit results for CDF and LHCb $\psi(nS)$ cross sections. It is apparent that the shape of the CDF $J/\psi$ production data is considerably better reproduced by this fit result than by the corresponding results of BK, as shown in Fig. 5.4, in the $p_T$ region that is included in the fit.

The outcome of this analysis is that $\psi(2S)$ production is completely dominated by the transverse ${}^3S_1^{[8]}$ channel, while in the case of direct $J/\psi$ production, the partial ${}^3S_1^{[8]}$ cross section is even negative, and rather large. The dominating processes, especially at high $p_T$, are the ${}^3P_J^{[8]}$ channels and, to a lesser extent, the ${}^1S_0^{[8]}$ channel. The direct production of $J/\psi$ and $\psi(2S)$ states is therefore surprisingly characterized, according to this analysis, by very different production mechanisms.

From the results of the fit they calculate the polarization observables. Figure 5.5 (bottom) shows the calculated polar anisotropy in the HX frame for the prompt $J/\psi$, which is only slightly transverse, and rather constant with $p_T$. The figure also shows the calculated polarizations of the feed-down contributions. The $\psi(2S)$ is calculated to be almost fully transverse at high $p_T$, increasing with $p_T$ from values that even reach slight longitudinal values at low $p_T$. The feed-down contribution of the $\chi_c$ states is calculated to contribute with a slightly transverse distribution to the direct $J/\psi$ component, which itself is also characterized by a slight transverse polarization.

**Table 5.1**  Summary of the $\psi(nS)$ LDME results of the individual NRQCD analyses

|  | BK [39, 45] | CMSWZ [43] | GWWZ [40] |
|---|---|---|---|
| $\mathcal{O}^{J/\psi}({}^1S_0^{[8]})$ [$10^{-2}$ GeV$^3$] | 4.97 | 8.9 | 9.7 |
| $\mathcal{O}^{J/\psi}({}^3S_1^{[8]})$ [$10^{-2}$ GeV$^3$] | 0.22 | 0.30 | −0.46 |
| $\mathcal{O}^{J/\psi}({}^3P_J^{[8]})$ [$10^{-2}$ GeV$^5$] | −1.61 | 1.26 | −2.14 |
| $\mathcal{O}^{\psi(2S)}({}^1S_0^{[8]})$ [$10^{-2}$ GeV$^3$] | −0.247 | – | −0.01 |
| $\mathcal{O}^{\psi(2S)}({}^3S_1^{[8]})$ [$10^{-2}$ GeV$^3$] | 0.280 | – | 0.34 |
| $\mathcal{O}^{\psi(2S)}({}^3P_J^{[8]})$ [$10^{-2}$ GeV$^5$] | 0.168 | – | 0.95 |

Additionally, the GWWZ group provides a NLO NRQCD analysis of the $\Upsilon(nS)$ states [41]. They take into account the feed-down contributions into the $\Upsilon(1S)$ and $\Upsilon(2S)$ states, but assume the $\Upsilon(3S)$ state to be free of feed-down decays.[3] However, due to the lack of $\chi_b$ hadroproduction data (at the time of this analysis) the LDMEs of the $\chi_{bJ}(nP)$ states cannot be directly obtained, and are mostly unconstrained parameters in the fit. The analysis uses hadroproduction data with $p_T > 8$ GeV from the CDF, LHCb, CMS and ATLAS experiments. The polarization measurements from CMS [35] and the updated CDF measurement [36] are considered in the fits. Again, an iterative procedure is chosen. First, the $\Upsilon(3S)$ data are fit and the corresponding LDMEs are fixed. Then, the $\Upsilon(2S)$ data are fit, fixing the $\chi_{bJ}(2P)$ and $\Upsilon(2S)$ LDMEs. Finally, the $\Upsilon(1S)$ data are fit, estimating the rest of the LDMEs.

The cross section data are nicely reproduced by these fit results. The polarization of the prompt $\Upsilon(nS)$ states is found to be almost fully transverse for the $\Upsilon(3S)$, dominated by ${}^3S_1^{[8]}$ production, but almost unpolarized for the $\Upsilon(1S)$ and $\Upsilon(2S)$, which are dominated by $\chi_b$ feed-down, which are characterized by a slight transverse polarization. The directly produced $\Upsilon(1S)$ is found to be dominated by the ${}^1S_0^{[8]}$ channel, as is the $\Upsilon(2S)$, to a lesser extent, with more significant ${}^3S_1^{[8]}$ and ${}^3P_J^{[8]}$ admixtures. The CMS polarization measurements can be reproduced by the calculations for the $\Upsilon(1, 2S)$, but on the other hand, the transverse $\Upsilon(3S)$ polarization is far off the experimental results, at high $p_T$.

## Summary of Existing NRQCD Analyses

Even though the NLO NRQCD calculations for the SDCs and polarizations of the individual color channels of the three groups are compatible, the conclusions from their individual NRQCD analyses are fundamentally different. The estimated composition of the individual production mechanisms is very different for each of the analyses, which can be appreciated by comparing the individual values of the LDMEs for $\psi(nS)$ production, as summarized in Table 5.1. These physically very different scenarios do not largely affect the inclusive cross section calculations, as with differ-

---

[3]The GWWZ group are in the progress of preparing an update of this analysis, including feed-down contributions of the decays $\chi_{bJ}(3P) \to \Upsilon(3S)$.

ent compositions very similar shapes of the $p_T$ distributions can be obtained. On the other hand, the calculations of the polarization observables are very different for the individual scenarios, which is illustrated in Fig. 5.6, comparing the outcome of the individual NRQCD analyses with the CMS measurements of S-wave quarkonium polarization. The differences are seen clearest when considering the J/$\psi$ calculations,

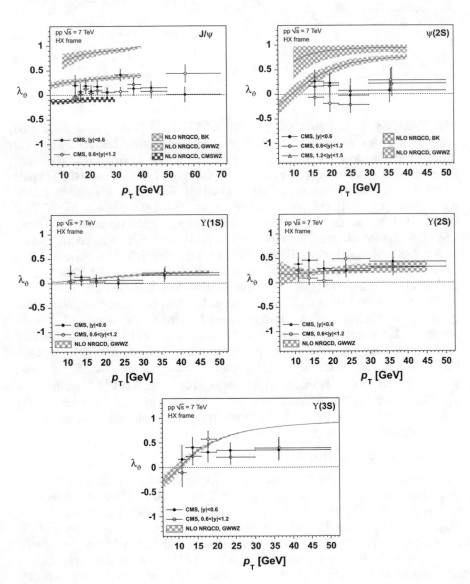

**Fig. 5.6** CMS measurements of the polarization parameter $\lambda_\vartheta^{HX}$ of prompt S-wave quarkonium states [31, 35], compared to NLO NRQCD calculations for direct production by BK [37, 45] (green), CMSWZ [43] (red) and for prompt production by GWWZ [40, 41] (cyan)

which range from slightly longitudinal polarization by CMSWZ, slightly transverse polarization by GWWZ, and almost fully transverse polarization by BK. How can these differences be understood, given that compatible theoretical inputs are used?

The main differences of the strategies and assumptions of the individual NRQCD analyses are summarized in Table 5.2. One main reason for the differences is the list of considered data points. While the choice of collision system does not seem to affect the results significantly, the $p_T$ requirements on the considered data points has a large influence on the results (see Sect. 5.3), as was observed already by BK in Ref. [50]. Furthermore, the decision to include the available polarization data in the fits is of great importance. One more major difference is the consideration of the feed-down decays, which has a large influence on the results, as shown in Refs. [40, 41]. Given the differences of the individual NLO NRQCD analyses and the resulting confusion in the field, it is advisable to reflect upon the assumptions made by the individual analyses, and their justification, in more detail.

While it is appreciated by all NRQCD analyses that a minimum $p_T$ requirement, $p_T^{min}$, is essential to ensure not being biased by the low-$p_T$ data in regions where the NLO NRQCD calculations cannot be trusted, due to non-perturbative effects, they nevertheless include rather low-$p_T$ data. The specific choices of $p_T^{min}$ are partially dictated by the available data. For example, the BK analysis would suffer from a loss of around 50 and 98.7% of the hadroproduction and photoproduction data, respectively, would they have chosen $p_T^{min} = 7\,\text{GeV}$. Given that no high-$p_T$ photoproduction data is available, they would have had to restrict the analysis to hadroproduction data, not being able to assess LDME universality. While the NRQCD analysis that is described in Sect. 5.3 finds that stable and unbiased results can only be obtained if data with $p_T^{min}/M_Q < 3$ is removed from the analysis [1], all NRQCD analyses described in this section use lower-$p_T$ data within the unstable region, possibly biasing the LDME results, especially the BK analysis, and the GWWZ bottomonium analysis. Given that the low-$p_T$ data are usually characterized by considerably smaller uncertainties with respect to the high-$p_T$ data, the choice to include the high-precision low-$p_T$ data can severely bias the analyses.

One general problem is the treatment of the uncertainties. None of the existing analyses include theoretical uncertainties in their fit programs, making it impos-

**Table 5.2** Summary of the main differences between various NRQCD analyses

| | BK [39, 45] | CMSWZ [43] | GWWZ [40, 41] |
|---|---|---|---|
| Collision systems considered | $pp$, $p\bar{p}$, $e^+e^-$, $ep$ | $p\bar{p}$ | $pp$, $p\bar{p}$ |
| Polarization data considered | No | Yes | No ($\psi(nS)$) Yes ($\Upsilon(nS)$) |
| $p_T$ region (GeV) | $p_T > 3$ ($pp$, $p\bar{p}$) $p_T > 1$ ($e^+e^-$, $ep$) | $p_T > 7$ | $p_T > 7$ ($\psi(nS)$) $p_T > 8$ ($\Upsilon(nS)$) |
| LDMEs estimated for | $\psi(nS)$ | $J/\psi$ | $\psi(nS)$, $\chi_{cJ}$ $\Upsilon(nS)$, $\chi_{bJ}(1, 2P)$ |
| Feed-down considered | No | No | Yes (except for $\Upsilon(3S)$) |

sible to interpret reasonably the minimized values of the normalized $\chi^2$. Another general point is the inconsistency of the treatment of the polarization-dependence of the cross section measurements within the fits. The measured quarkonium cross sections depend on an assumption on the polarization of the quarkonium state, due to acceptance effects of the experiments. The general practice followed by all LHC experiments is to publish quarkonium cross sections for the unpolarized assumption, as well as additional information that allows one to recalculate the cross sections for any given assumption on the polarization. This so-called "polarization envelope" can be rather large, as can be seen for example in the ATLAS $\Upsilon(nS)$ analysis [12], where the cross section results vary by up to $+353$ and $-65\%$ for low-$p_T$ results, and still by up to $+50$ and $-13\%$ for results with $p_T > 30\,\text{GeV}$, for different polarization scenarios. While some of the NRQCD analyses calculate large polarizations, especially at high $p_T$, in the fits these analyses compare their models to the cross section measurements that assume unpolarized production. This inconsistent treatment of the cross section measurements should be removed from the analyses.

Another important point is the treatment of the feed-down contributions. It is clear that not including the feed-down components can significantly bias the result, making it impossible to reasonably interpret the results. On the other hand, including the feed-down components in situations where there is no experimental information available about the relative importance of these decays, the results can be misleading. This is, for example, the case in the bottomonium analysis of GWWZ, where they include the feed-down decays $\chi_{bJ}(2P) \rightarrow \Upsilon(1, 2S) + \gamma$, without any experimental constraints on the size of this effect. Given that the corresponding LDMEs are free in the fit, unconstrained due to the non-existent $\chi_{bJ}(2P)$ data, the fit has excessive freedom which may result in a good description of the data while not reflecting reality. A specific example where this freedom has possibly led to a bias is the feed-down fraction of $\chi_{bJ}(2P) \rightarrow \Upsilon(2S)+\gamma$, which is reported to be 35–76% for hadroproduction, increasing with $p_T$ [41], which seems to be significantly overestimated, considering the corresponding results from LHCb, especially at high $p_T$ [29].

Contrary to the cross section measurements, the polarization observables allow for an immediate interpretation of the results, given that the polarization calculations of the individual color channels are very different. This power of the available polarization measurements has not been exploited by several of the NRQCD analyses. Given the observed inconsistencies at the Tevatron, it is a reasonable choice to exclude the pre-LHC polarization results from the analyses. However, with the new results from polarization analyses emerging from the LHC, it is imperative to move these measurements to the center of the attention, in future analyses.

These NRQCD analyses were all by-products of the daunting calculations of the NLO NRQCD calculations, for the SDCs and polarizations, for various collision systems, CM energies, states and kinematic regions. The emphasis of the work was certainly put on a rigorous calculation of the inputs, while no rigorous strategies for the comparison of these theoretical inputs to the data were developed. This led to the introduction of several inconsistencies and simplifications in these analyses, such as neglecting the polarization-dependence of the cross section measurements, not considering global experimental and theoretical uncertainties, not considering feed-down decays and polarization data. In addition, in these analyses low-$p_T$ data was used, which may bias the results, as will be shown in the next section.

While the remarkable effort of calculating these NLO NRQCD inputs cannot be overestimated, the corresponding comparisons to data led to large differences between the individual calculations, all labelled as "NLO NRQCD", giving the wrong impression that NLO NRQCD is either very unstable concerning small differences in the theoretical inputs, or cannot be trusted to provide reliable quantitative predictions at all.

## 5.3 A Data-Driven Perspective

The current status of the understanding of quarkonium production with the NRQCD factorization approach, as discussed in the previous section, is not satisfactory, given that the various NLO NRQCD results, based on compatible theoretical inputs, mutually exclude each other. The problems of the individual NRQCD analyses motivate a more rigorous phenomenological study, avoiding said problems, to clarify quarkonium production. This analysis has been conducted in collaboration with Pietro Faccioli, Carlos Lourenço, João Seixas and Hermine Wöhri, resulting in the publication [1], on which this section is almost entirely based, providing further detail.

The analysis strategy is guided by two main data-driven observations, which allow us to infer certain statements about the composition of the individual color channels, even before employing a fit of the LDMEs.

1. The $p_T/M_\mathcal{Q}$ scaling observed in the LHC quarkonium cross section data suggests that all quarkonium states are produced in a very similar way, likely dominated by one CO mechanism (see Sect. 5.1.1).
2. The finding of the LHC experiments that none of the S-wave quarkonium states shows any strong polarization suggests that this dominating CO contribution is the unpolarized $^1S_0^{[8]}$ term (see Sect. 5.1.2).

These interpretations clearly follow the Occam's razor principle. To study the validity of these hypotheses, a sophisticated fitting method is developed, utilizing MCMC methods to estimate the LDMEs. Theoretical uncertainties are taken into account in the fitting procedure, correlated between the individual quarkonium states, as well as correlated uncertainties from the experimental results, such as uncertainties originating from the luminosity measurements. Moreover, the polarization-dependence of the cross section measurements is fully taken into account.

The polarization observables are moved to the center of the attention, conducting a true "global fit" analysis of quarkonium production, to profit from the now available high-quality LHC quarkonium polarization data. All $\psi(2S)$ and $\Upsilon(3S)$ cross section and polarization data as measured by the LHC experiments, available at the time of the submission of the publication, are considered in this analysis. The analysis is restricted to the highest-mass S-wave states of the charmonium and bottomonium system, to avoid having to take into account the feed-down decays of heavier quarkonium states. While the assumption that the $\psi(2S)$ measurements allow access to exclusive direct production is considered very safe, the assumption that the $\Upsilon(3S)$ state is free of significant feed-down contributions seems to have been invalidated by the recent LHCb measurement [29], finding relatively large feed-down fractions from the $\chi_b(3P)$ states. This information was not available at the time of completion of this analysis. Therefore, the $\Upsilon(3S)$ results of this analysis have to be interpreted carefully, and will be updated in a forthcoming publication.

The main idea of this study is to search for a "kinematic domain of validity" of the NLO NRQCD calculations, at high $p_T$, where all quarkonium hadroproduction data from the LHC experiments can be well described by the calculations, which is achieved by a "kinematic domain scan", consecutively removing low-$p_T$ data, until a stable region of the fit is reached.

First, the theory ingredients are introduced in Sect. 5.3.1, before discussing the fitting method in detail in Sect. 5.3.2. The results of the kinematic domain scan are discussed in Sect. 5.3.3, and the results obtained from data within the found safety domain, as well as corresponding predictions, are shown in Sect. 5.3.4. A brief discussion of the main differences of the analysis strategy and physics conclusions with respect to the other existing NRQCD analyses in Sect. 5.3.5, is followed by an interpretation of the results in Sect. 5.3.6.

### 5.3.1   Theory Ingredients

State-of-the-art NLO calculations of the NRQCD SDCs and polarizations of the CS and the individual CO channels, for $pp$ hadroproduction at $\sqrt{s} = 7$ TeV, from Refs. [37, 45] are used for this analysis. A subset of these calculations, at mid-rapidity, has been shown in Fig. 2.5.

In order to be able to cover as much phase space as possible, to ensure that all available data can be included in the study, some straightforward extrapolations of the theoretical inputs are performed, extending the models towards lower $p_T$ for mid-rapidity calculations ($|y| < 1.2$), and filling the gap in rapidity between the mid-rapidity and forward-rapidity calculations, which are explicitly performed for the LHCb phase space $2 < y < 4.5$. The $p_T$ shapes of the calculations are found not to depend significantly on $y$. However, the normalization slightly decreases as a function of $|y|$, slightly differently for the individual color channels, a feature taken into account in these extrapolations.

These calculations are provided for a rest energy of the $Q\bar{Q}$ pair of $E_0 = 3$ GeV. In order to obtain the shape and the polarizations for a quarkonium state with mass $M_Q$, the transverse momentum $p_T$ of the calculations is rescaled according to $p'_T = p_T \cdot \frac{M_Q}{E_0}$. Even though the actual rest energy of the $Q\bar{Q}$ pair, $E_{Q\bar{Q}}$ does not necessarily have to coincide with $M_Q$, the rescaling used here is a very good approximation of the quarkonium production kinematics, as shown in Ref. [1].

The normalization of the SDCs is rescaled as well, with an exponential function of the quarkonium mass, whose exponent is determined from cross section measurements at the LHC, from normalization factors corresponding to the ones used to construct the individual functions in Fig. 5.1. The normalization scale is defined by $N(M_Q) = a \cdot \exp(-b \cdot M_Q)$, with $a = 16.82$ and $b = 0.94\,\text{GeV}^{-1}$, leading to $N(M^{\psi(2S)}) = 0.53$ and $N(M^{\Upsilon(3S)}) = 1.0 \cdot 10^{-3}$. Given that any change in normalization of the SDCs can be compensated by changes in the numerical values of the LDMEs, except for the CS contribution, which is negligible, the conclusions of this study are not affected by this normalization scaling, albeit affecting the values of the LDMEs, for instance for the $\psi(2S)$, by a factor of around 2. This redefinition of the LDMEs has to be considered when comparing their numerical values to those of other NRQCD analyses. However, this definition ensures that the values of the LDMEs obtained for different quarkonium states can be very easily interpreted. If, for two different states, the LDMEs of a given color channel, $\mathcal{O}^Q(n)$, are identical, the normalization scaling ensures that the probabilities of the transition of a $Q\bar{Q}$ pair in state $n$ into the two quarkonium states are identical. This also means, thanks to the $p_T/M_Q$ scaling, that the relative importance of the partial cross sections, $\sigma(n)$, with respect to the full quarkonium cross section, $\sigma(Q)$, is identical for the two states, at any given value of $p_T/M_Q$.

**Construction of the Theoretical Uncertainties**

As shown in Fig. 2.5 (top right), the LO and NLO calculation for the SDCs and polarization observables show large differences, especially for the CS $^3S_1^{[1]}$ and the CO $^3P_J^{[8]}$ channels. It is imperative to associate a theoretical uncertainty to this fact, implemented in the fitting procedure. Assuming that the perturbative expansion in powers of $\alpha_s$ is convergent, one can expect that the "true values" are contained within the range [NLO-$\Delta$, NLO+$\Delta$], with $\Delta = $ NLO-LO, the difference between the NLO and LO calculations. Therefore, in this fitting method, the theoretical inputs are characterized by a Gaussian probability distribution, centered at the NLO values, with a width corresponding to $\Delta$. Given that in the case of the CS calculation more information is available in the literature, the corresponding probability distribution of the CS polarization calculations is modeled by a Gaussian probability distribution, centered at the NLO values, but with a width corresponding to $\Delta^* = $ NNLO* $-$ NLO, the difference between the NNLO* calculations [51, 52] and the NLO values. Considering the large differences between the NLO and NNLO* calculations for the CS SDC, an asymmetric Gaussian is used, centered at NLO, with a width of $\Delta$ for SDC values below the NLO values, and with a width $\Delta^*$ for SDC values above the NLO values, in order to constrain the CS SDCs to positive values.

In the fitting method, each color channel is associated with a nuisance parameter (NUP), which is allowed to float within the Gaussian constraint. On the one hand, this allows the fit to adjust the shape of the theoretical models within the boundaries defined by the Gaussian probability distributions, and on the other hand, this procedure ensures that the output of the fit reflects the theoretical uncertainties. Furthermore, the NUPs of the individual color channels for different quarkonium states are correlated in a simultaneous fit, ensuring that the individual models are adjusted in the same way for the individual states. If, for example, the fit prefers a CS contribution close to the NNLO* calculations rather than the NLO values, this is the case for both the $\psi(2S)$ and $\Upsilon(3S)$ states.

## The Peculiar $^3P_J^{[8]}$ Component

The changes from LO to NLO are especially large for the $^3P_J^{[8]}$ component, with the polar anisotropy changing from close to 0 to values beyond +1 (at high $p_T$), and the SDC k-factors reaching large values of up to 250 at high $p_T$ and even becoming negative at low $p_T$ [45]. Large k-factors by themselves do not necessarily change the physics conclusions of the analyses, as normalization changes can always be compensated by the fitted LDMEs. However, in this case the relevant and alarming property is the huge change of the k-factors as a function of $p_T$, strongly affecting the shape. This results in a huge theoretical uncertainty that induces instabilities in the fit, and can make the corresponding fit results meaningless, as the fit is dominated by the freedom of the shape of the $^3P_J^{[8]}$ component.

Besides the large changes from LO to NLO, the $^3P_J^{[8]}$ NLO calculations themselves reveal some peculiarities. The SDC changes sign at approximately 7.5 GeV, being positive at low $p_T$ and negative at high $p_T$. The partial cross section of the $^3P_J^{[8]}$ can of course be made positive at high $p_T$, if the corresponding LDME is found to be negative, but it remains true that negative partial $^3P_J^{[8]}$ cross sections cannot be avoided for both the low- and high-$p_T$ regions, except for the special case where the LDME is 0. In addition, the polar anisotropy $\lambda_\vartheta^{HX}$ also reflects an unphysical behavior. From low to high $p_T$, several phases can be observed, starting from around 0, passing through unphysical regions below $-1$, diverging to $-\infty$, changing sign, and then passing through unphysical values beyond +1, approaching +1 at high $p_T$. The unphysical behavior of the $^3P_J^{[8]}$ term suggests that the NLO calculations cannot be trusted, and that large higher-order corrections should be expected.

These unphysical properties of the $^3P_J^{[8]}$ term are usually not regarded as a conceptual problem. It is argued that only the color-inclusive observables can be measured experimentally and interpreted physically. However, it is clear that negative partial cross sections that vary strongly as a function of $p_T$ require a compensation by positive partial cross sections of the other components, ensuring that the overall cross section and polarization is physical, in all kinematic regions, including the angular dimensions. This compensation is especially delicate when the $^3P_J^{[8]}$ channel is a dominating contribution. The LDME of the $^3P_J^{[8]}$ component has to be small enough to ensure that the sum with the remaining contributions leads to a positive cross section in all kinematic regions, and to an angular distribution characterized by anisotropy parameters satisfying the constraints of the allowed phase space regions (see Fig. 2.11).

This reasoning is of course only valid if the individual color channels cannot be observed individually. As is argued in Ref. [1], the properties of the individual color channels, their SDCs and polarization observables at NLO, are sufficiently different, making it possible to distinguish, on a statistical basis, the individual components, by defining kinematic discriminants, including, for example, the angular variables

$\cos \vartheta$ and $\varphi$. Due to the transverse angular distribution of the $^3P_J^{[8]}$ component, its contribution is enhanced with respect to differently polarized contributions at values close to $|\cos \vartheta^{HX}| = 1$, and reduced close to $|\cos \vartheta^{HX}| = 0$. With respect to an unpolarized contribution, this enhancement can reach a factor of 1.5. With respect to a longitudinal contribution, this enhancement can even reach $+\infty$, corresponding to the possibility of a complete and unambiguous separation of an unphysical contribution, at least in an ideal "Gedankenexperiment". As a simplified pedagogical example [53], one can construct a hypothetical quarkonium production scenario as a composition of an unpolarized $^1S_0^{[8]}$ component with a partial cross section of 55 nb, and a transversely polarized $^3P_J^{[8]}$ component ($\lambda_\vartheta^{HX} = +1$) with a negative partial cross section of $-45$ nb, leading to a total quarkonium cross section of 10 nb. Applying a selection criterion in the experimental setup of $|\cos \vartheta^{HX}| > 0.8$, the observed quarkonium cross section would be negative, $-1.24$ nb, obviously not an example that can be realized in nature. This example should illustrate how delicate the compensation of the positively contributing color channels has to be, in the case of the presence of a channel with negative partial cross section and/or unphysical polarization.

In this analysis, the hypothesis that all contributing processes are individually observable, and therefore required to be physical, is explored. This corresponds to the very intuitive view that quarkonium production can be separated in the two phases described by NRQCD factorization, with an observable intermediate object, the $Q\bar{Q}$ pair, where both phases, the production of the $Q\bar{Q}$ pair, as well as its hadronization into the quarkonium state, are observable and measurable processes. Due to the reasons discussed in this section, the $^3P_J^{[8]}$ component is neglected in this analysis. However, it is included in the fits a posteriori, testing the effect of this term, which is found to be negligible (see Sect. 5.3.3).

## 5.3.2  Fitting Method

Instead of the parametrization of quarkonium cross sections and polarizations with LDMEs, according to Eqs. 2.1 and 2.2, this analysis makes use of so-called "fraction parameters" (FPs). The relative importance of a given color channel with respect to the full quarkonium cross section, at a reference point $p_T^*/M_Q$, is denoted as the fraction $f(n) = \frac{\sigma^*(n)}{\sigma^*(Q)}$, and the relative importance of a given CO channel with respect to the sum of all partial CO cross sections, $\sigma_{co}^*(Q) = \sum_{n \neq cs} \sigma^*(n)$, is denoted as the "octet-fraction" $f_{co}(n) = \frac{\sigma^*(n)}{\sigma_{co}^*(Q)}$, at the same reference point. The full quarkonium cross section at this reference point is of course given as $\sigma^*(Q) = \sigma_{cs}^*(Q) + \sigma_{co}^*(Q)$ with $\sigma_{cs}^*(Q)$ the CS partial cross section. Finally, the "octet-ratio" $R_{co}$ is defined as the ratio of the sum of the CO partial cross sections with respect to the CS partial cross section, $R_{co} = \sigma_{co}^*(Q)/\sigma_{cs}^*(Q)$. The vector $\vec{f} = (R_{co}, f_{co}(n_1), \ldots, f_{co}(n_i))$ summarizes the FPs, with the constraint $\sum_{n \neq cs} f_{co}(n) = 1$. Contrary to the intuitive notion that any fraction should be contained between 0 and 1, the octet-fraction of a given CO component can exceed 1, in case this is compensated by a negative octet-fraction of a different component.

With these definitions of $\vec{f}$, the NRQCD factorization theorem, as introduced in Eq. 2.1, can be written as

$$\sigma(Q) = \sigma_{cs}(Q) + R_{co} \cdot \sigma_{cs}^*(Q) \cdot \sum_{n \neq cs} S[Q\bar{Q}] \frac{f_{co}(n)}{S^*[Q\bar{Q}(n)]} , \qquad (5.2)$$

where $\sigma_{cs}(Q)$ and $S[Q\bar{Q}(n)]$ depend on the quarkonium kinematics, while the parameters labeled with $^{(*)}$ are constants that can be determined univocally from the SDC calculations, at the reference point $p_T^*/M_Q$, and the CS LDME. For this analysis, the arbitrary reference point is chosen to be $p_T^*/M_Q = 6$. In this notation, the free LDME parameters have been substituted by free $\vec{f}$ parameters. The values of the octet-fractions and the octet-ratio can be directly transformed into LDMEs and vice versa. However, the fraction-notation of the quarkonium cross section is more intuitive to comprehend. Most importantly, from a technical point of view, finding the optimal values of the fractions is more straightforward than those of the LDMEs, as the numerical values of the fractions and their uncertainties are more easily constrainable in the fitting method. Therefore, the fit makes use of the fraction parameters, which are then, a posteriori, transformed directly to the LDMEs, the parameters of interest (POI) of this analysis.

The fitting method is based on a least $\chi^2$ approach. The total $\chi^2$ is calculated as the sum of the partial $\chi^2$ values from each considered data point $i$,

$$\chi_i^2 = \left( \frac{x_i^{\text{data}} - x_i^{\text{model}}}{\sigma_i^{\text{data}}} \right)^2 , \qquad (5.3)$$

with $x_i^{\text{data}}$ the central value and $\sigma_i^{\text{data}}$ the total uncertainty of the measurement, and $x_i^{\text{model}}$ the model calculation, for the current value of $\vec{f}$ and the nuisance parameters. The observable $x$ represents either a cross section measurement $\sigma_i^{\text{data}}(Q)$, or a measurement of the polar anisotropy in the HX frame, $\lambda_{\vartheta,i}^{\text{data}}$. In the case of a cross section measurement, the model $\sigma_i^{\text{model}}(Q)$ is calculated by summing the partial cross sections of the individual channels, for the given kinematic region of the measurement, according to Eq. 5.2. In the case of a polarization measurement, $\lambda_{\vartheta,i}^{\text{model}}$ is calculated by combining the polarizations of the individual channels according to Eq. 2.6, depending on the relative importance of the individual channels given by the current values of $\vec{f}$, and the kinematic region of the measurement.

As for the CMS quarkonium polarization analyses, a MCMC approach is chosen to sample the total $\chi^2$ function, in order to obtain the full posterior probability density function, in a multi-dimensional form for all POI, FPs and NUPs. The approach is implemented in the same way as the MCMC method discussed in detail in Sect. 4.1.3, relying on a Metropolis–Hastings criterion, with adapted widths of the proposal functions for the burn in period. The prior distribution is chosen to be uniform in the fraction parameters. The approach has been validated by an independent implementation of the MIGRAD minimization algorithm of the MINUIT package [54].

Central values and confidence intervals of the POI can be constructed from this multi-dimensional PPD by projections on the individual parameters, as described in Sect. 4.1.4. The obtained NRQCD models can be visualized by error bands that can be directly built from the full information of the PPD, including the variations according to all POI and NUPs. This method ensures that correlations between the parameters are taken into account and can be studied in detail.

**Nuisance Parameters**

Several NUPs are floating in the fit, taking into account various effects. The theory uncertainties are, as described above, implemented as NUPs that are allowed to float within Gaussian constraints, changing the values of $x_i^{\text{model}}$ in the calculation of the partial $\chi_i^2$. A simultaneous fit of the $\psi(2S)$ and $\Upsilon(3S)$ data ensures that the NUPs associated with the theoretical uncertainties affect the models of the individual states in the same way.

Global uncertainties of the measurements, affecting all data points of a given set of measurements, such as uncertainties on the luminosity estimation in the case of cross section measurements, are also implemented as NUPs, floating within a Gaussian constraint. The data points $x_i^{\text{data}}$ are scaled synchronously by these global uncertainties, in the same way for all data points belonging to the same data-taking period of a given experiment, correlating these individual sets of measurements.

The Gaussian constraints of the NUPs are enforced by adding a penalty term to the total $\chi^2$, increasing with the difference of the current value of the NUP with respect to the center of the Gaussian constraint, normalized by its width.

The dependence of the experimental cross section measurements on the polarization is taken into account in the following way. For each set of the POI, FPs and NUPs, sampled by the MCMC algorithm, both the cross section model $\sigma_i^{\text{model}}(\mathcal{Q})$ and the associated polarization model $\lambda_{\vartheta,i}^{\text{model}}$ are calculated. The measured value of the cross section $\sigma_i^{\text{data}}(\mathcal{Q})$ can then be scaled according to the current calculated polarization scenario, by utilizing the published information from the experimental acceptance effects for this given measurement.

These implementations ensure that the correlations of the individual uncertainties, both theoretical and experimental, are taken into account correctly, and further ensure that the full information about the cross section measurements, including their polarization dependence, is correctly taken into account, avoiding the inconsistent treatment followed in previous NRQCD analyses.

## 5.3.3  Kinematic Domain Scan

It is known since the birth of NRQCD that the low-$p_{\text{T}}$ region cannot be expected to be accurately described by perturbative NRQCD calculations, due to non-perturbative effects, affecting the validity of the double expansion in $\alpha_s$ and $v^2$. Previous analyses have chosen to define a fixed (arbitrary) value for the low-$p_T$ cutoff $p_T^{min}$. This analysis attempts to define the domain of validity of the NLO NRQCD calculations in a rigorous way, by conducting a kinematic domain scan, continuously increasing the cutoff value $p_T^{min}$, until a stable region of the fit is reached.

For this scan, a total of 121 data points is used, measured in $pp$ collisions at 7 TeV by three LHC experiments: CMS polarization and cross section data of the $\psi(2S)$ [3, 31] and $\Upsilon(3S)$ [8, 35], LHCb cross section data of the $\psi(2S)$ [14] and $\Upsilon(3S)$ [15], as well as ATLAS cross section data of the $\Upsilon(3S)$ [12]. The lowest $p_T^{min}$ considered in this analysis is $p_T^{min} = 4$ GeV for the $\psi(2S)$ data and $p_T^{min} = 10$ GeV for the $\Upsilon(3S)$ data. This leads to a total of 43 data points for the $\psi(2S)$ and 78 data points for the $\Upsilon(3S)$, with 99 cross section measurements, and 22 measured values of $\lambda_\vartheta^{HX}$.

To ensure a stable kinematic domain scan, no theoretical uncertainties are considered in this step of the analysis. Therefore, the $\psi(2S)$ and $\Upsilon(3S)$ data scans can be regarded as uncorrelated, and are conducted separately. As mentioned above, the $^3P_J^{[8]}$ component is not considered in this scan, the LDMEs $\mathcal{O}^{\psi(2S)}(^3S_1^{[8]})$, $\mathcal{O}^{\psi(2S)}(^1S_0^{[8]})$ of the $\psi(2S)$ state, and the LDMEs $\mathcal{O}^{\Upsilon(3S)}(^3S_1^{[8]})$, $\mathcal{O}^{\Upsilon(3S)}(^1S_0^{[8]})$ of the $\Upsilon(3S)$ state are the POI of the individual fits, and their behavior is studied as a function of the cutoff value $p_T^{min}$.

Figure 5.7 shows all relevant diagrams describing the results of the kinematic domain scan, as a function of $p_T^{min}$ or $p_T^{min}/M_Q$: The reduced minimized $\chi^2$ (top) is studied as a function of $p_T^{min}/M_Q$, in order to identify a stable region of the fit, where the data are well described. The fit quality is incredibly bad when including low-$p_T$ data, characterized by $\chi^2$-probabilities of the order of $10^{-55}$, mostly driven by the transversely polarized models due to a dominant $^3S_1^{[8]}$ contribution, incompatible with the considered data. With increasing $p_T^{min}$, the relative importance of the $^1S_0^{[8]}$ contribution increases, leading to smaller polarizations, and the fit quality improves, resulting in reasonable $\chi^2$-probabilities of the order of a few 10%, affected by some fluctuations. The absolute values of the reduced $\chi^2$ do not necessarily have to be reasonable, as for example, theoretical uncertainties are not considered in these fits. However, a stabilization of the fit can be identified at the point where the reduced $\chi^2$ values stop their decreasing trend, and start fluctuating around a constant value. As can be seen in the top panel, this is the case for $p_T^{min}/M_Q$ values of around 3, very similar for the $\psi(2S)$ and $\Upsilon(3S)$. This chosen value of $p_T^{min}/M_Q = 3$ corresponds to roughly $p_T^{min} = 11$ GeV for the $\psi(2S)$ and $p_T^{min} = 31$ GeV for the $\Upsilon(3S)$.

The fit results change dramatically as a function of $p_T^{min}$, and therefore the possible physics conclusions of different NRQCD analyses with different values of $p_T^{min}$. The left panel in the middle row shows the behaviors of the LDMEs of the $^3S_1^{[8]}$ and $^1S_0^{[8]}$ terms, for the $\psi(2S)$ scan. For low values of $p_T^{min}$, the $^3S_1^{[8]}$ LDME is large, and decreases significantly when excluding low-$p_T$ data. Obviously, the trend of the $^1S_0^{[8]}$ term is opposite, its value increasing steeply as a function of $p_T^{min}$. This behavior is emphasized in the bottom left panel, which shows the behavior of the ratio of the $^3S_1^{[8]}$ and $^1S_0^{[8]}$ LDMEs, which drops by two orders of magnitude in the considered region of $p_T^{min}$. The right panel in the middle row shows the behaviors of the LDMEs of the $^3S_1^{[8]}$ and $^1S_0^{[8]}$ terms for the $\Upsilon(3S)$ scan. The same behavior as in the $\psi(2S)$

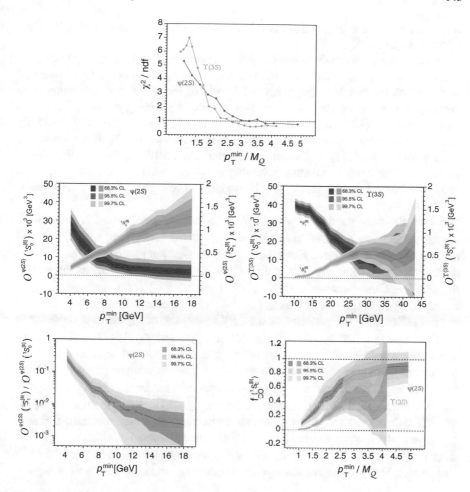

**Fig. 5.7** Results of the kinematic domain scan, as a function of $p_T^{min}$ or $p_T^{min}/M_Q$: reduced minimized $\chi^2$ of the $\psi(2S)$ and $\Upsilon(3S)$ scans (*top*), the LDMEs $\mathcal{O}^{\psi(2S)}(^3S_1^{[8]})$ and $\mathcal{O}^{\psi(2S)}(^1S_0^{[8]})$ (*middle row left*), the LDMEs $\mathcal{O}^{\Upsilon(3S)}(^3S_1^{[8]})$ and $\mathcal{O}^{\Upsilon(3S)}(^1S_0^{[8]})$ (*middle row right*), the ratio of the LDMEs $\mathcal{O}^{\psi(2S)}(^3S_1^{[8]})/\mathcal{O}^{\psi(2S)}(^1S_0^{[8]})$ (*bottom left*) and the octet-fraction $f_{co}(^1S_0^{[8]})$ of the $^1S_0^{[8]}$ term of the $\psi(2S)$ and $\Upsilon(3S)$ scans, at the chosen reference point $p_T^*/M_Q = 6$ (*bottom right*) [1]

case is observed, with the $^3S_1^{[8]}$ LDME decreasing and the $^1S_0^{[8]}$ LDME increasing as a function of $p_T^{min}$. However, in this case the effect is not as dramatic as in the $\psi(2S)$ case. For the understanding of the actual mixture of the individual color channels it is advisable to look at the behavior of the octet-fractions $f_{co}(^1S_0^{[8]})$ of the $^1S_0^{[8]}$ term, at the chosen reference point of $p_T^*/M_Q = 6$, as shown in the bottom right panel of Fig. 5.7, increasing for the $\psi(2S)$ from around $f_{co}(^1S_0^{[8]}) = 10$–90% and for the $\Upsilon(3S)$ from $f_{co}(^1S_0^{[8]}) = 1$% to around 50%. This diagram clearly shows that in the validity domain, at high $p_T/M_Q$, the partial cross section of the $^1S_0^{[8]}$ is almost fully dominating for the $\psi(2S)$, and is a considerable contribution also in the $\Upsilon(3S)$ case, of the order of 50%.

In conclusion, this analysis has identified a domain of validity of NLO NRQCD calculations, as the kinematic region of around $p_T/M_Q > 3$. As can be appreciated from the middle panels of Fig. 5.7, the LDMEs of the $\Upsilon(3S)$ are stabilizing once entering the domain of validity. However, carefully studying the behavior of the LDMEs of the $\psi(2S)$ it might be the case that the LDMEs stabilize only at slightly higher values of $p_T/M_Q$, which can only be probed once the now available higher-$p_T$ cross section data [5, 11] are included in the analysis. Therefore, the specific value of $p_T/M_Q > 3$ defining the domain of validity is not claimed to be the final numerical value valid for all quarkonia, and will be updated once higher precision data will be included.

Given the very good fit qualities when requesting $p_T/M_Q > 3$, these studies do not indicate the need to include the neglected $^3P_J^{[8]}$ term, with $\mathcal{O}^Q(^3P_J^{[8]})$ defined to be 0 in the scans. Nevertheless, the data-driven assumption that one can neglect the $^3P_J^{[8]}$ contribution is addressed at this point of the analysis, by adding the additional free parameters $\mathcal{O}^{\psi(2S)}(^3P_J^{[8]})$ and $\mathcal{O}^{\Upsilon(3S)}(^3P_J^{[8]})$, and repeating the fit requiring $p_T/M_Q > 3$. In the case of the $\psi(2S)$, the central values of the result are not affected by the presence of the $^3P_J^{[8]}$ term, the octet-fractions resulting in $f_{co}(^1S_0^{[8]}) = (80 \pm 8)\%$, $f_{co}(^3S_1^{[8]}) = (20 \pm 20)\%$ and $f_{co}(^3P_J^{[8]}) = (0 \pm 20)\%$, consistent with the results obtained in the kinematic domain scan, and with the hypothesis that the $^3P_J^{[8]}$ term can be neglected. The uncertainties on the $^3P_J^{[8]}$ and $^3S_1^{[8]}$ are large in the fit, and strongly anti-correlated, as can be appreciated in Fig. 5.8, showing the correlations of the octet-fractions $f_{co}(^3S_1^{[8]})$ and $f_{co}(^3P_J^{[8]})$ in the $\psi(2S)$ fit. In the case of the $\Upsilon(3S)$ fit, the system becomes strongly under-constrained, inducing large correlations between all free LDMEs, nevertheless favoring the $^1S_0^{[8]}$ term with $f_{co}(^1S_0^{[8]}) = 80^{+70}_{-30}\%$. These results support the decision to neglect the $^3P_J^{[8]}$ component, and indicate that this hypothesis does not induce a bias in the analysis.

### 5.3.4  Results and Predictions

Within the identified domain of validity, the analysis can now be repeated, profiting from the full developed machinery, including the theoretical uncertainties, simultaneously considering the $\psi(2S)$ and $\Upsilon(3S)$ data. In the validity domain, 44 data points remain, 30 cross section measurements and 14 measurements of $\lambda_\vartheta^{HX}$. Figure 5.9 shows the fitted data points and the corresponding models, including the color-inclusive model and the individual color channels, represented by their central curves and their uncertainties. The reduced $\chi^2$ of the simultaneous fit of the $\psi(2S)$ and $\Upsilon(3S)$ data is 36.2/40, and the corresponding $\chi^2$-probability is 64%, representing a very good fit quality.

This nominal fit is characterized by 4 POI, the LDMEs $\mathcal{O}^{\psi(2S)}(^3S_1^{[8]})$, $\mathcal{O}^{\psi(2S)}(^1S_0^{[8]})$ of the $\psi(2S)$ state, and the LDMEs $\mathcal{O}^{\Upsilon(3S)}(^3S_1^{[8]})$, $\mathcal{O}^{\Upsilon(3S)}(^1S_0^{[8]})$ of the $\Upsilon(3S)$ state.

**Fig. 5.8** Correlations in the PPD of the octet-fractions $f_{CO}(^3S_1^{[8]})$ and $f_{CO}(^3P_J^{[8]})$ of the $\psi(2S)$ fit including the $^3P_J^{[8]}$ component, at the chosen $p_T^*/M_Q$ reference point [1]

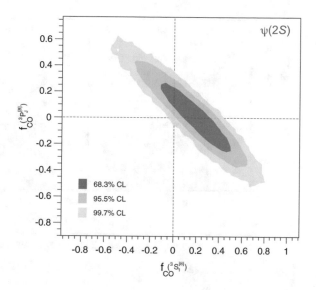

Furthermore, the fit contains 3 NUPs describing the theoretical uncertainties, one for each color channel, and 4 NUPs describing the global uncertainties associated to 4 different data-taking periods (ATLAS, LHCb and CMS 2010, plus CMS 2011 data). While the fit prefers a small CS contribution close to NLO-$\Delta$, the $^1S_0^{[8]}$ term is preferred to be close to NLO+$\Delta$, and the $^3S_1^{[8]}$ term is preferred to be close to the NLO calculations. These preferences correspond to the best-fit values, but they are not constrained well enough by the data, with large uncertainties on the corresponding NUPs, to draw definite conclusions regarding the full calculations of the SDCs and polarizations of the individual channels.

This nominal fit results in an octet-fraction of around $f_{CO}(^1S_0^{[8]}) = 83 \pm 10\%$ and an octet-ratio of around $R_{co} = 12 \pm 2$ for the $\psi(2S)$, and in $f_{CO}(^1S_0^{[8]}) = 51 \pm 13\%$ and $R_{co} = 11 \pm 2$ for the $\Upsilon(3S)$. The large values of the octet-ratios $R_{co}$ show that the relative importance of the CS contribution with respect to the full cross section is below 10%, at the reference point. The corresponding results of the POI are obtained from the almost symmetric and close-to-Gaussian 1-dimensional projections of the PPD:

$$\mathcal{O}^{\psi(2S)}(^3S_1^{[8]}) = 1.0 \pm 0.7 \cdot 10^{-4} \text{ GeV}^3 \,,$$

$$\mathcal{O}^{\psi(2S)}(^1S_0^{[8]}) = 2.2 \pm 0.3 \cdot 10^{-2} \text{ GeV}^3 \,,$$

$$\mathcal{O}^{\Upsilon(3S)}(^3S_1^{[8]}) = 3.6 \pm 1.5 \cdot 10^{-4} \text{ GeV}^3 \,,$$

$$\mathcal{O}^{\Upsilon(3S)}(^1S_0^{[8]}) = 1.5 \pm 0.2 \cdot 10^{-2} \text{ GeV}^3 \,.$$

Figure 5.10 (left) shows the 2-dimensional projections of the PPDs on $\mathcal{O}^Q(^3S_1^{[8]})$ and $\mathcal{O}^Q(^1S_0^{[8]})$, for both states, showing that the LDMEs of the $\psi(2S)$ and $\Upsilon(3S)$

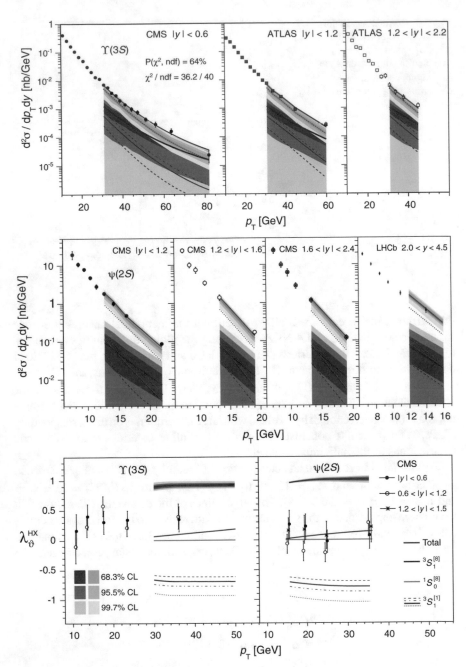

**Fig. 5.9** Data points considered in the nominal fit, compared to the color-inclusive model (*blue*) as well as to the individual color channels, represented by their central curves and their uncertainties. The CS contribution is represented by the LO (*dashed*), NLO (*dot-dashed*) and NNLO* (*dotted*) calculations, as well as by the corresponding best-fit curve (*solid*) [1]

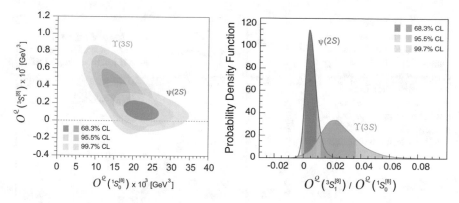

**Fig. 5.10**  2-dimensional projections of the PPDs on $\mathcal{O}^{\mathcal{Q}}(^3S_1^{[8]})$ and $\mathcal{O}^{\mathcal{Q}}(^1S_0^{[8]})$, for both $\psi(2S)$ and $\Upsilon(3S)$ states (*left*), and 1-dimensional projections of the PPD on the $\mathcal{O}^{\mathcal{Q}}(^3S_1^{[8]})$ over $\mathcal{O}^{\mathcal{Q}}(^1S_0^{[8]})$ ratio, for both $\psi(2S)$ and $\Upsilon(3S)$ states (*right*) [1]

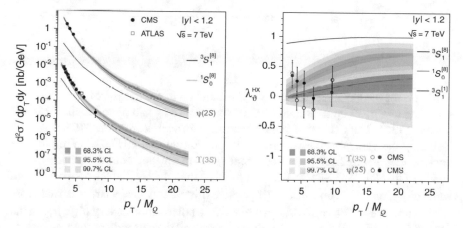

**Fig. 5.11**  Predictions for the $p_T$-differential cross section (*left*) and polarization (*right*) observables, as a function of $p_T/M_{\mathcal{Q}}$, valid in the kinematic region $|y| < 1.2$. The data points considered in the global fit are also shown [1]

are very similar, compatible with being the same, although the fit suggests that the octet-fraction $f_{co}(^1S_0^{[8]})$ is larger in the $\psi(2S)$ case, with a larger admixture of the $^3S_1^{[8]}$ channel in the $\Upsilon(3S)$ case. Due to the $p_T/M_{\mathcal{Q}}$ and normalization scalings, this means that the octet-fractions, and therefore the relative importance and the mixture of the individual physical processes, are compatible with being identical for the two states, at any value of $p_T/M_{\mathcal{Q}}$. Figure 5.10 (right) shows 1-dimensional projections of the PPD on the $\mathcal{O}^{\mathcal{Q}}(^3S_1^{[8]})$ over $\mathcal{O}^{\mathcal{Q}}(^1S_0^{[8]})$ ratio, for both $\psi(2S)$ and $\Upsilon(3S)$ states. The ratio of the LDMEs is compatible for both states, and is rather small, below 1.5% for the $\psi(2S)$ and below 6% for the $\Upsilon(3S)$, at 95% CL.

From these fit results, one can calculate predictions for the cross section and polarization observables, towards higher $p_T/M_{\mathcal{Q}}$ values. These predictions are shown in

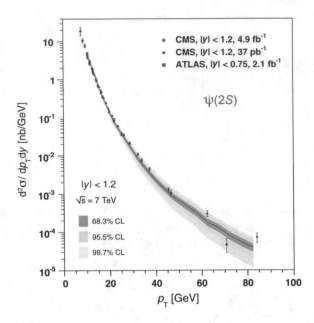

**Fig. 5.12** Predictions for the $p_T$-differential cross section for the $\psi(2S)$ state [1], in the kinematic region $|y| < 1.2$, compared to high-$p_T$ $\psi(2S)$ cross section data from ATLAS [11] and CMS [5], not included in the global fit, as well as to low-$p_T$ CMS results [3] that are included in the fit

Fig. 5.11, for the kinematic region $|y| < 1.2$ and up to around $p_T/M_Q = 22$, corresponding to a $p_T$ of up to around $80\,\text{GeV}$ for the $\psi(2S)$, and up to around $225\,\text{GeV}$ for the $\Upsilon(3S)$. The CL bands exploit the full information from the PPD, therefore including the variations corresponding to the POI and all NUPs, representing the uncertainties arising from both theoretical and experimental contributions.

Given that the $^3S_1^{[8]}$ SDC is flatter at high $p_T$ than the $^1S_0^{[8]}$ SDC, above a certain value of $p_T$, the $^3S_1^{[8]}$ contribution (if positive) necessarily gets a larger partial cross section than the $^1S_0^{[8]}$ term, leading to transverse polarization at very high $p_T$. In the case of the $\psi(2S)$, this "turnover" point is far beyond the currently probed $p_T$ region. However, given the larger $^3S_1^{[8]}/^1S_0^{[8]}$ ratio in the $\Upsilon(3S)$ case, the corresponding turnover point is pushed towards a lower value of $p_T/M_Q$, as can be seen in Fig. 5.11 (left), only slightly beyond the region currently probed by the experiments.

The predictions can be compared to $\psi(2S)$ production cross sections measured by the ATLAS [11] and CMS [5] Collaborations, with a $p_T$ reach of 70 and $80\,\text{GeV}$, respectively, made public only recently. These results are not included in the global fit procedure of this analysis. Given that the $p_T$ reach of the cross section data points included in the global fit does not exceed $22\,\text{GeV}$, the comparison of these predictions with the newly available high-$p_T$ $\psi(2S)$ data is very interesting. This comparison is shown in Fig. 5.12, also including the CMS results [3] that are included in the global fit. The high-$p_T$ results are well compatible with the prediction of this analysis. This is quite remarkable, considering the limited kinematic region of the cross section measurements that enter the global fit. This finding strongly supports the validity of this study.

### 5.3.5  Comparison with Other NRQCD Analyses

To put the results of this analysis into context, the strategy of this analysis can be compared to those of the other NRQCD analyses (see details in Sect. 5.2 and Table 5.2). The most important differences concern the data considered for the global fits. Certainly, the consideration of the polarization data is an important point. Also the addition of the photoproduction data in the case of BK has to be mentioned as possible source of difference. Another important difference is the $p_T$ region considered in the analyses. Considering that fairly different values of $p_T^{min}$ have been used, ranging from 1 up to 8 GeV in the previous analyses, compared to the domain of validity identified by this analysis, no compatible results can be expected to be obtained, as was shown in Sect. 5.3.3. Besides these clear differences, the treatment of the polarization-dependence of the cross section measurement, the treatment of the experimental and theoretical uncertainties, and the $p_T/M_Q$ as well as the normalization scaling represent differences that have to be considered. In addition, the $^3P_J^{[8]}$ component was included in all previous analyses, contrary to this analysis.

Due to the $p_T/M_Q$ and normalization scalings, the numerical values of the LDMEs cannot be directly compared to the values obtained by the other groups. However, redoing the full analysis, without the scalings, including data with $p_T > 3$ GeV, the results for the $\psi(2S)$ LDMEs reproduce those obtained by the BK analysis (see Table 5.1). Furthermore, their LDMEs can be reproduced with or without including the polarization constraints in the global fit. These tests indicate that differences between this analysis and the one of BK do not originate from the inclusion of photoproduction data nor from neglecting the $^3P_J^{[8]}$ term (which is a small negative contribution in the case of the BK $\psi(2S)$ result), but are solely based on the different value of $p_T^{min}$, and possibly the treatment of the theoretical and experimental uncertainties.

The ratios of the LDMEs are not sensitive to the $p_T/M_Q$ and normalization scalings, so that the ratios obtained in this analysis (see the right panel of Fig. 5.10) can be directly compared to the ratios obtained by BK and GWWZ. The comparison can be done for the $\psi(2S)$ results of the BK and GWWZ analyses (see Table 5.1), and for the $\Upsilon(3S)$ results of the GWWZ analysis [41]. For all those analyses, the $^1S_0^{[8]}$ component is negligible, even slightly negative, the same being true for the $^3P_J^{[8]}$ component. Therefore, the $^3S_1^{[8]}$ component completely dominates the production in these analyses, leading to transverse polarization in all cases. The conclusions regarding the relative importance of the individual color channels for quarkonium production are therefore very different from those of this analysis. This difference can be quantified by looking at the ratios $\mathcal{O}^Q(^3S_1^{[8]})/\mathcal{O}^Q(^1S_0^{[8]})$ of the individual analyses. The ratios are $-1.1$ and $-34$ for the BK and GWWZ $\psi(2S)$ results, respectively, and $-2.5$ for the GWWZ $\Upsilon(3S)$ result. Compared to the result of this analysis, below 1.5% for the $\psi(2S)$ and below 6% for the $\Upsilon(3S)$, at 95% CL, the ratios of the previous analyses point to a completely different mixture of processes, albeit affected by large and not fully quantified uncertainties.

Obviously, as shown in Fig. 5.6, a dominance of the transversely polarized $^3S_1^{[8]}$ term cannot explain the experimental observations, while this analysis, as shown in Fig. 5.9, describes very well the experimental results, for both quarkonium production cross sections as well as quarkonium polarizations.

### 5.3.6  Conclusions

NRQCD, as a rigorous, consistent and effective theory based on full QCD, is expected to describe observables of quarkonium production accurately. Within this model, the non-perturbative parameters of the theory, the LDMEs, are expected to follow certain velocity scaling rules, leading to the prediction that, for the production of S-wave quarkonia, the $^1S_0^{[8]}$, $^3S_1^{[8]}$ and $^3P_J^{[8]}$ LDMEs should be of the same order of magnitude, as they all scale with $v^4$ (see Table 2.1). However, this study suggests the existence of stronger hierarchies, as the data indicate that the LDMEs follow the hierarchy $^1S_0^{[8]} \gg {}^3S_1^{[8]} \gg {}^3P_J^{[8]}$. The $\psi(2S)$ and $\Upsilon(3S)$ quarkonium states seem to be preferentially produced through transitions involving $Q\bar{Q}$ pairs characterized by low spin and angular momentum, while the transitions via $Q\bar{Q}$ pairs in higher angular momentum or spin states seem to be suppressed. These new hierarchies seen in the LHC data can be explained by building conjectures about the dynamics of quarkonium bound state formation. An example of such a conjecture is attempted in [1].

It is interesting to note that the observed suppression of the $^3S_1^{[8]}$ term with respect to the $^1S_0^{[8]}$ term is smaller for the $\Upsilon(3S)$ than for the $\psi(2S)$, indicating slight differences between the production of charmonium and bottomonium states, however, still compatible with having the same mixture of production processes, to be studied in more detail once higher precision data is available. However, when comparing the results of the $\psi(2S)$ analysis with those of the $\Upsilon(3S)$ analysis, it has to be emphasized again that the $\chi_b(3P) \rightarrow \Upsilon(3S)$ feed-down effect is not taken into account.

Following the guidance of the first data-driven observation, leading to the suspicion that all quarkonia, both S-wave and P-wave, are produced from similar underlying production mechanisms, it is reasonable to assume that the observed hierarchies in the $\psi(2S)$ and $\Upsilon(3S)$ data are also realized in the production of the other quarkonium states, to be confirmed. The production of P-wave states is expected to be dominated by $^3S_1^{[8]}$ CO production, given that the $^1S_0^{[8]}$ channel is assumed to be suppressed with respect to the $^3S_1^{[8]}$ channel by a factor of $v^4$ (see Table 2.1). Therefore, in current NRQCD analyses of P-wave states, usually only the $^3S_1^{[8]}$ channel is considered. However, given the central finding of this analysis, that the $^3S_1^{[8]}$ channel is suppressed with respect to the $^1S_0^{[8]}$ channel, future analyses of P-wave states should include the $^1S_0^{[8]}$ term as well. The measurement of the polarizations of the P-wave states at the LHC will eventually provide crucial inputs to distinguish between the P-wave production via $^3S_1^{[8]}$ and $^1S_0^{[8]}$ intermediate states.

The finding that NLO NRQCD can only be reliably used to calculate observables for high-$p_T$ regions may seem disappointing. However, considering that the $^1S_0^{[8]}$

component is unpolarized in all orders in $\alpha_s$, and the experimental observation that also for low-$p_T/M_Q$ values all S-wave quarkonia are unpolarized, just as in the high-$p_T/M_Q$ region, allows for the possibility that quarkonium production in the low-$p_T/M_Q$ region can be explained by the same dominant production channel, the $^1S_0^{[8]}$. The data themselves do not suggest a change in production mechanisms. Therefore, the factorization approach may still hold at low $p_T$, and improved calculations of the SDCs at higher orders may be able to describe consistently quarkonium production down to lower-$p_T$ values than currently possible with the available NLO calculations.

One basic ingredient of NRQCD is LDME universality, requiring that the transition of an initial $Q\bar{Q}$ pair in a certain state $n$ has the same probability to hadronize into a bound quarkonium state, independently of the mechanism that produced the initial $Q\bar{Q}$ state, may it be through $pp$, $e^+e^-$, $ep$, or heavy-ion collisions, or even through Higgs decays such as $H \to Q^{3S_1} + \gamma$ [55]. This important prediction cannot be tested, given that all currently available photoproduction data belong to the kinematic domain that is excluded by this analysis. Therefore, LDME universality can only be tested once high-$p_T$ quarkonium production data are available for collision systems other than $pp$ and $p\bar{p}$, or alternatively, through measurements involving associated production of quarkonium states [56].

This analysis constitutes a rigorous study of the LHC data in terms of NLO NRQCD calculations, by identifying a kinematic domain of validity, for which the NLO NRQCD calculations can be trusted, and can very well describe the $\psi(2S)$ and $\Upsilon(3S)$ polarization data as well as the corresponding $p_T$-differential cross sections. The results indicate $^1S_0^{[8]}$-octet dominance, solving the problem of the slowly-converging perturbative expansion in NRQCD, as the dominantly contributing channel is only affected by very small differences between the LO and NLO calculations. If the dominance of $^1S_0^{[8]}$ CO production is confirmed for other S-wave and P-wave states, in line with the data-driven observations, with extended global fits including feed-down decays, these studies will provide vital information towards the understanding of hadron formation, complementing further theoretical developments [57–59].

## 5.4 Results Summary

This chapter has summarized the impact of the measurements of quarkonium production cross sections and polarizations at the LHC experiments, in the quest to understand how quarks bind into quarkonia. The new high-quality quarkonium polarization data, as well as the cross section measurements, extending the reach towards higher $p_T$, allow for two data-driven observations, qualitatively addressing quarkonium production. The LHC cross section measurements suggest that all quarkonia are produced in a very similar way, dominated by one color-octet production channel. Furthermore, the LHC quarkonium polarization results suggest that this dominating channel is the unpolarized $^1S_0^{[8]}$ term, given that the results cluster around the unpolarized limit.

NRQCD provides state-of-the-art calculations for cross section and polarization observables in $pp$ collisions at the LHC, up to next-to-leading order. Given that the non-perturbative part of quarkonium production cannot be calculated, but is contained fully within the supposedly universal LDMEs, data has to be used to constrain these parameters. Several NRQCD analyses exist, using a multitude of different strategies and assumptions, resulting in calculations for polarization and cross section observables that mutually exclude each other.

An independent and original phenomenological analysis of LHC $\psi(2S)$ and $\Upsilon(3S)$ production results was developed, inspired and guided by data-driven observations, moving the polarization observables to the center of the study. Problems of previous analyses are identified and avoided. A systematic search for the domain of validity of next-to-leading order NRQCD is successfully performed, suggesting that current calculations are not reliable for the region $p_T/M_Q < 3$. The data in the remaining domain of validity can be nicely described with the surprising result that the unpolarized $^1S_0^{[8]}$ color-octet channel dominates quarkonium production, providing a straightforward solution to the long-standing "quarkonium polarization puzzle".

These observations should be confirmed by extended global fits of quarkonium data, including all S-wave and P-wave states. Moreover, extended measurements at the LHC are needed to constrain the LDMEs of the P-wave states. Especially important is the measurement of the polarization of the P-wave states, as this will allow us to differentiate between the $^3S_1^{[8]}$ and $^1S_0^{[8]}$ intermediate $Q\bar{Q}$ states.

Contrary to the status-quo in the pre-LHC era, this chapter suggests, based on experimental measurements at the LHC and new phenomenological interpretations, that direct quarkonium production proceeds preferentially through initial quark-antiquark pairs which are in the unpolarized $^1S_0^{[8]}$ configuration, revealing new hierarchies of the LDMEs with respect to the NRQCD velocity scaling rules.

# References

1. Faccioli P, Knünz V, Lourenço C, Seixas J, Wühri H (2014) Quarkonium production in the LHC era: a polarized perspective. Phys Lett B 736:98
2. CMS Collaboration (2011) Prompt and non-prompt $J/\psi$ production in pp collisions at $\sqrt{s} = 7$ TeV. Eur Phys J C 71:1575
3. CMS Collaboration (2012) $J/\psi$ and $\psi(2S)$ production in $pp$ collisions at $\sqrt{s} = 7$ TeV. J High Energy Phys 1202:011
4. CMS Collaboration (2014) $J/\psi$ and $\psi(2S)$ prompt double-differential cross sections in $pp$ collisions at 7 TeV. CMS-PAS-BPH-14-001
5. CMS Collaboration (2015) Measurement of prompt $J/\psi$ and $\psi(2S)$ double-differential cross sections in $pp$ collisions at $\sqrt{s} = 7$ TeV. Submitted to Phys Rev Lett. arXiv:1502.04155
6. CMS Collaboration (2011) Upsilon production cross-section in pp collisions at $\sqrt{s} = 7$ TeV. Phys Rev D 83:112004
7. CMS Collaboration (2013) Measurement of the $\Upsilon(1S)$, $\Upsilon(2S)$, and $\Upsilon(3S)$ cross sections in pp collisions at $\sqrt{s} = 7$ TeV. Phys Lett B 727:101
8. CMS Collaboration (2013) $\Upsilon(1S)$, $\Upsilon(2S)$ and $\Upsilon(3S)$ cross section measurements in $pp$ collisions at $\sqrt{s} = 7$ TeV. CMS-PAS-BPH-12-006

9. CMS Collaboration (2015) Measurements of the $\Upsilon(1S)$, $\Upsilon(2S)$ and $\Upsilon(3S)$ differential cross sections in $pp$ collisions at $\sqrt{s} = 7$ TeV. Submitted to Phys Lett B. arXiv:1501.07750

10. ATLAS Collaboration (2011) Measurement of the differential cross-sections of inclusive, prompt and non-prompt $J/\psi$ production in proton-proton collisions at $\sqrt{s} = 7$ TeV. Nucl Phys B 850:387

11. ATLAS Collaboration (2014) Measurement of the production cross-section of $\psi(2S) \rightarrow J/\psi(\rightarrow \mu^+\mu^-)\pi^+\pi^-$ in pp collisions at $\sqrt{s} = 7$ TeV at ATLAS. J High Energy Phys 1409:79

12. ATLAS Collaboration (2013) Measurement of Upsilon production in 7 TeV pp collisions at ATLAS. Phys Rev D 87:052004

13. LHCb Collaboration (2011) Measurement of $J/\psi$ production in $pp$ collisions at $\sqrt{s} = 7$ TeV. Eur Phys J C 71:1645

14. LHCb Collaboration (2012) Measurement of $\psi(2S)$ meson production in pp collisions at $\sqrt{s} = 7$ TeV. Eur Phys J C 72:2100

15. LHCb Collaboration (2012) Measurement of Upsilon production in pp collisions at $\sqrt{s} = 7$ TeV. Eur Phys J C 72:2025

16. LHCb Collaboration (2013) Production of $J/\psi$ and $\Upsilon$ mesons in $pp$ collisions at $\sqrt{s} = 8$ TeV. J High Energy Phys 1306:064

17. ALICE Collaboration (2011) Rapidity and transverse momentum dependence of inclusive J/$\psi$ production in $pp$ collisions at $\sqrt{s} = 7$ TeV. Phys Lett B 704:442

18. ALICE Collaboration (2012) Measurement of prompt $J/\psi$ and beauty hadron production cross sections at mid-rapidity in $pp$ collisions at $\sqrt{s} = 7$ TeV. J High Energy Phys 1211:065

19. CMS Collaboration (2012) Measurement of the relative prompt production rate of $\chi_{c2}$ and $\chi_{c1}$ in $pp$ collisions at $\sqrt{s} = 7$ TeV. Eur Phys J C 72:2251

20. CMS Collaboration (2013) Measurement of the $\sigma(\chi_{b2}(1P))/\sigma(\chi_{b1}(1P))$ production cross section ratio in pp collisions at $\sqrt{s} = 8$ TeV. CMS-PAS-BPH-13-005

21. CMS Collaboration (2014) Measurement of the production cross section ratio $\sigma(\chi_{b2}(1P))/\sigma(\chi_{b1}(1P))$ in $pp$ collisions at $\sqrt{s} = 8$ TeV. Submitted to Phys Lett B. arXiv:1409.5761

22. ATLAS Collaboration (2012) Observation of a new $\chi_b$ state in radiative transitions to $\Upsilon(1S)$ and $\Upsilon(2S)$ at ATLAS. Phys Rev Lett 108:152001

23. ATLAS Collaboration (2014) Measurement of $\chi_{c1}$ and $\chi_{c2}$ production with $\sqrt{s} = 7$ TeV $pp$ collisions at ATLAS. J High Energy Phys 1407:154

24. LHCb Collaboration (2012) Measurement of the cross-section ratio $\sigma(\chi_{c2})/\sigma(\chi_{c1})$ for prompt $\chi_c$ production at $\sqrt{s} = 7$ TeV. Phys Lett B 714:215

25. LHCb Collaboration (2013) Measurement of the relative rate of prompt $\chi_{c0}$, $\chi_{c1}$ and $\chi_{c2}$ production at $\sqrt{s} = 7$ TeV. J High Energy Phys 10:115

26. LHCb Collaboration (2014) Measurement of the $\chi_b(3P)$ mass and of the relative rate of $\chi_{b1}(1P)$ and $\chi_{b2}(1P)$ production. J High Energy Phys 1410:88

27. LHCb Collaboration (2012) Measurement of the ratio of prompt $\chi_c$ to $J/\psi$ production in $pp$ collisions at $\sqrt{s} = 7$ TeV. Phys Lett B 718:431

28. LHCb Collaboration (2012) Measurement of the fraction of $\Upsilon(1S)$ originating from $\chi_b(1P)$ decays in $pp$ collisions at $\sqrt{s} = 7$ TeV. J High Energy Phys 11:031

29. LHCb Collaboration (2014) Study of $\chi_b$ meson production in $pp$ collisions at $\sqrt{s} = 7$ and 8 TeV and observation of the decay $\chi_b$ (3P) $\rightarrow \Upsilon$ (3S)$\gamma$. Eur Phys J C 74:3092

30. HERA-B Collaboration (2009) Kinematic distributions and nuclear effects of $J/\psi$ production in 920-GeV xed-target proton-nucleus collisions. Eur Phys J C 60:525

31. CMS Collaboration (2013) Measurement of the prompt $J/\psi$ and $\psi(2S)$ polarizations in pp collisions at $\sqrt{s} = 7$ TeV. Phys Lett B 727:381

32. LHCb Collaboration (2013) Measurement of $J/\psi$ polarization in $pp$ collisions at $\sqrt{s} = 7$ TeV. Eur Phys J C 73:2631

33. LHCb Collaboration (2014) Measurement of $\psi(2S)$ polarisation in $pp$ collisions at $\sqrt{s} = 7$ TeV. Eur Phys J C 74:2872

34. ALICE Collaboration (2012) $J/\psi$ polarization in $pp$ collisions at $\sqrt{s} = 7$ TeV. Phys Rev Lett 108:082001

35. CMS Collaboration (2013) Measurement of the $\Upsilon(1S)$, $\Upsilon(2S)$ and $\Upsilon(3S)$ polarizations in $pp$ collisions at $\sqrt{s} = 7$ TeV. Phys Rev Lett 110:081802
36. CDF Collaboration (2012) Measurements of angular distributions of muons from $\Upsilon$ meson decays in $p\bar{p}$ collisions at $\sqrt{s} = 1 : 96$ TeV. Phys Rev Lett 108:151802
37. Butenschön M, Kniehl B (2012) $J/\psi$ polarization at Tevatron and LHC: nonrelativistic-QCD factorization at the crossroads. Phys Rev Lett 108:172002
38. Butenschön M, Kniehl B (2011) Reconciling $J/\psi$ production at HERA, RHIC, Tevatron, and LHC with NRQCD factorization at next-to-leading order. Phys Rev Lett 106:022003
39. Butenschön M, Kniehl B (2011) World data of $J/\psi$ production consolidate NRQCD factorization at NLO. Phys Rev D 84:051501
40. Gong B, Wan LP, Wang JX, Zhang HF (2013) Polarization for prompt $J/\psi$, $\psi(2S)$ production at the Tevatron and LHC. Phys Rev Lett 110:042002
41. Gong B, Wan LP, Wang JX, Zhang HF (2014) Complete next-to-leading-order study on the yield and polarization of $\Upsilon(1S, 2S, 3S)$ at the Tevatron and LHC. Phys Rev Lett 112:032001
42. Ma YQ, Wang K, Chao KT (2011) $J/\psi$ $(\psi')$ production at the Tevatron and LHC at $\mathcal{O}(\alpha_s^4 v^4)$ in nonrelativistic QCD. Phys Rev Lett 106:042002
43. Chao KT, Ma YQ, Shao HS, Wang K, Zhang YJ (2012) $J/\psi$ polarization at hadron colliders in nonrelativistic QCD. Phys Rev Lett 108:242004
44. CDF Collaboration (1997) $J/\psi$ and $\psi(2S)$ production in $p\bar{p}$ collisions at $\sqrt{s} = 1.8$ TeV. Phys Rev Lett 79:572
45. Private Communication from M. Butenschön and B. Kniehl
46. CDF Collaboration (2000) Measurement of $J/\psi$ and $\psi(2S)$ polarization in $p\bar{p}$ collisions at $\sqrt{s} = 1 : 8$ TeV. Phys Rev Lett 85:2886
47. CDF Collaboration (2007) Polarization of $J/\psi$ and $\psi(2S)$ mesons produced in $p\bar{p}$ collisions at $\sqrt{s} = 1 : 96$ TeV. Phys Rev Lett 99:132001
48. Faccioli P, Seixas J (2012) Observation of $\chi_c$ and $\chi_b$ nuclear suppression via dilepton polarization measurements. Phys Rev D 85:074005
49. Han H, Ma YQ, Meng C, Shao HS, Zhang YJ, Chao KT (2014) $\Upsilon(nS)$ and $\Upsilon(nP)$ production at hadron colliders in nonrelativistic QCD. *1410.8537*, 2014
50. Butenschön M, Kniehl B (2012) $J/\psi$ production in NRQCD: a global analysis of yield and polarization. Nucl Phys B-Proc Suppl 151:222–224
51. Artoisenet P, Campbell JM, Lansberg JP et al (2008) $\Upsilon$ production at fermilab Tevatron and LHC energies. Phys Rev Lett 101:152001
52. Lansberg JP (2009) Real next-to-next-to-leading-order QCD corrections to $J/\psi$ and Upsilon hadroproduction in association with a photon. Phys Lett B 679:340
53. Private Communication from P. Faccioli
54. James F, Winkler M (2004) Minuit User's Guide. CERN, Geneva. http://seal.web.cern.ch/seal/documents/minuit/mnusersguide.pdf
55. Bodwin GT, Petriello F, Stoynev S, Velasco M (2013) Higgs boson decays to quarkonia and the $H\bar{c}c$ coupling. Phys Rev D 88:053003
56. QWG Collaboration (2011) Heavy quarkonium: progress, puzzles, and opportunities. Eur Phys J C 71:1534
57. Kang ZB, Qiu JW, Sterman G (2012) Heavy quarkonium production and polarization. Phys Rev Lett 108:102002
58. Kang ZB, Ma YQ, Qiu JW, Sterman G (2014) Heavy quarkonium production at collider energies (I): factorization and evolution. YITP-SB-13-48
59. Bodwin GT, Chung HS, Kim UR, Lee J (2014) Fragmentation contributions to $J/\psi$ production at the Tevatron and the LHC. Phys Rev Lett 113:022001

# Chapter 6
# Conclusions

## 6.1 Thesis Summary

The quest to understand hadron formation within the non-perturbative sector of QCD has led to a large activity in the field of research studying quarkonium production observables, both on the theoretical and experimental sides. Quarkonia are ideal objects to study how quarks bind into hadrons. Detailed studies of quarkonium production allow us, in a way that no other process does, to understand the interaction dynamics involving the long-distance strong force, and to ultimately understand how quarks bind into hadrons via the strong interaction. The theoretical treatment of quarkonium production is facilitated by the large masses of the charm and beauty quarks, seemingly allowing a factorization of quarkonium production into the production of an initial $Q\bar{Q}$ intermediate "pre-resonant" state, and the subsequent non-perturbative evolution into the bound quarkonium system, characterized by two distinct time scales. This conjectured separation is exploited by the NRQCD factorization approach, which performs calculations of quarkonium production observables as a superposition of color-singlet and color-octet intermediate $Q\bar{Q}$ states, and is generally regarded as the most promising technique to understand quarkonium production through a QCD-inspired model.

The exhilarating success of NRQCD, after describing the Tevatron cross section measurements, was followed by disappointing predictions of quarkonium polarization observables, calculated to be transverse, as these were found to be very different from the Tevatron data. However, the Tevatron quarkonium polarization results were ambiguous and inconsistent, diminishing the impact of this problem. In order to clarify these issues, the mission for the LHC-era was clearly stated as a twofold strategy, on the one hand providing a clear and unambiguous experimental picture of quarkonium polarization, and on the other hand providing reliable and global phenomenological interpretations of quarkonium production measurements within the framework of NRQCD (and possibly beyond).

The LHC can be regarded as quarkonium factory, thanks to its high center-of-mass energies and luminosities. Among the LHC experiments, the CMS experiment

© Springer International Publishing Switzerland 2017
V. Knünz, *Measurement of Quarkonium Polarization to Probe QCD at the LHC*, Springer Theses, DOI 10.1007/978-3-319-49935-2_6

is especially well suited to study quarkonium production cross sections and polarization observables, up to very high $p_T$. Indeed, CMS has performed a wealth of measurements of quarkonium cross sections and polarizations, albeit mostly constrained to measurements of S-wave quarkonium states. Quarkonium cross sections were measured by all LHC experiments, extending the kinematic reach considerably. Thanks to the excellent performance of the CMS detector, its muon momentum and vertexing resolutions, and its efficient trigger system, the quarkonium polarization parameters of all S-wave states could be measured with low statistical and systematic uncertainties, extending the $p_T$ reach substantially with respect to previous measurements. A detailed description of these polarization measurements constitute the first core topic of this thesis. Other LHC experiments have also measured quarkonium polarization at low $p_T$, complemented by updated Tevatron results on this topic, leading to the very pleasant situation that the LHC era quarkonium polarization measurements are all consistent, throughout all kinematic regions and experiments, with the somewhat surprising result that all measurements are clustering around the unpolarized limit. This consistent experimental picture is an important step towards the understanding of quarkonium production, given that now the clean polarization constraints can be used in the global interpretation of quarkonium production results, which was not reasonably possible before the LHC, due to experimental inconsistencies.

Several groups attempted an interpretation of the LHC quarkonium production results, in next-to-leading order NRQCD analyses, with the goal to estimate the values of the long-distance matrix elements, and therefore identify the dominant production channels. Although starting from compatible theoretical calculations, the results of these analyses are very different, mutually excluding each other. These differences are caused by the various assumptions and different strategies that the individual groups chose in order to constrain the LDMEs. One common problem of all previous analyses is that low-$p_T$ data are included in their fits, which cannot be expected to be reliably described by perturbative NRQCD calculations. Following these NRQCD analyses, an independent and original phenomenological study was performed, constituting the second core topic of this thesis. This analysis is guided by the observation of two general features of the data. First, the similarity of the LHC quarkonium cross section measurements of various S-wave and P-wave states suggest that all quarkonia are produced very similarly, likely dominated by one color-octet contribution only. The second observation originates from the quarkonium polarization measurements, which do not show any significant polarization, throughout all S-wave quarkonium states and kinematic regions. This suggests that the dominant color-octet channel has to be the unpolarized $^1S_0^{[8]}$ intermediate state. This NRQCD analysis treats in the correct way all experimental and theoretical uncertainties, including, for the first time, the polarization-dependence of the experimental cross section results. The main idea of this analysis is to perform a kinematic domain scan, considering all $\psi(2S)$ and $\Upsilon(3S)$ LHC data, removing systematically low-$p_T$ data points, to find a domain of validity of next-to-leading order NRQCD in which the fits are stable, and the data can be well described. This domain of validity is found to be in the region $p_T/M_{\mathcal{Q}} > 3$. Within this stable region, this analysis leads to a coherent picture of quarkonium production cross sections and polarizations, providing a straightforward solution to the "quarkonium polarization puzzle", confirming the data-driven conjecture that the unpolarized intermediate $^1S_0^{[8]}$ $Q\bar{Q}$ state dominates quarkonium production, for the $\psi(2S)$ and $\Upsilon(3S)$ states. These findings provide vital information for the understanding of the fundamental processes leading to bound state formation in QCD.

## 6.2 Outlook

Although much progress has been made in the understanding of quarkonium production in the last few years, as summarized above, several open issues remain to be understood theoretically, and have to be confirmed with higher precision experimentally.

The cross section of the S-wave states have been measured for all states, up to relatively large values of $p_T$. The experimentally more difficult to access P-wave states have partially been measured at the LHC experiments. Nevertheless, more measurements have to be performed in order to understand the $\chi_c$ and $\chi_b$ production cross sections, and the feed-down fractions into the S-wave states, covering the highest possible $p_T$ regions. The polarizations measured for the S-wave states provide vital information, but they need to be extended towards higher $p_T$, and with higher precision than currently available, especially for the $\psi(2S)$ and $\Upsilon(3S)$ states. Given the experimental difficulties for the reconstruction of low-energy photons, no measurements of P-wave polarizations are available on this date. These measurements are especially important, given that they will allow us to distinguish between the $^3S_1^{[8]}$ and $^1S_0^{[8]}$ intermediate $Q\bar{Q}$ states in the case of P-wave production. It is imperative to test the suppression of the $^3S_1^{[8]}$ intermediate state also for P-wave quarkonia.

Already with today's available data it is possible to extend the data-driven NRQCD analysis as described in this thesis to the full charmonium system, including the feed-down decays of the heavier charmonium states, with the goal to estimate the LDMEs of the $\psi(nS)$ and $\chi_{cJ}$ states, providing predictions for cross sections and polarizations of all charmonium states, up to high $p_T$. A similar global analysis in the bottomonium system is certainly possible, but statistically limited, due to the lack of available data. The ultimate goal of these NRQCD analyses relying on LHC $pp$ data is to estimate the LDMEs of all S-wave and P-wave quarkonium states, in order to test today's conjecture that all quarkonia are dominantly produced through $^1S_0^{[8]}$ transitions. It will be very interesting to see if charmonium and bottomonium states show significant differences, due to the different heavy quark masses. Further theory developments in this sector, not discussed in this thesis, will in parallel try to gain information about the fundamental processes leading to quarkonium production.

One major point, to be addressed in the future, is the test of LDME universality. The validity of this basic ingredient of NRQCD can only be tested through measurements from collision systems different from $pp$, or through measurements involving associated production of quarkonium states. Testing universality is very important in order to enhance the predictive power of NRQCD in any other quarkonium production channels, such as in heavy-ion collisions, and for future measurements of the $Hc\bar{c}$ and $Hb\bar{b}$ couplings through the decays $H \to \mathcal{Q}^{3S_1} + \gamma$.

With the restart of $pp$ collisions at the LHC in 2015, with a center-of-mass energy of 13 TeV and an increased instantaneous luminosity, quarkonium production yields are expected to increase by almost a factor of four with respect to previous running conditions, increasing the $p_T$ reach of the measurements. With the upgraded CMS detector, and newly developed dimuon trigger strategies and algorithms, it will be possible to continue to lead the field of quarkonium production measurements among the LHC experiments, providing the results needed to draw definitive conclusions regarding the new findings concerning additional hierarchies within NRQCD, as found in the LHC Run I data. An intriguing era in quarkonium production physics has been concluded with the analysis of LHC Run I data, leading to vital information regarding the understanding of the fundamental non-perturbative processes leading to QCD bound state formation, to be further tested with data collected in LHC Run II.

# Curriculum Vitae

**Valentin Knünz**

valentinknuenz@gmail.com, +43 650 7674059
Vinzenzgasse 16/6, 1180 Vienna, Austria

## Research Experience

| | |
|---|---|
| 10/2015 − present | **Research Fellow at CERN Experimental Physics Dep., Geneva**<br>*Analytical research physicist within the CMS experiment at the LHC, CERN*<br>° Coordination of analysis working group (>50 contr. from ≈20 universities)<br>° Analysis of CMS data, focussing on the parametrization of angular distr.<br>° Phenomenological interpretation of LHC data<br>° Member of analysis review committees, peer review activities<br>° Particle detectors R&D, silicon pixel module testing and production |
| 04/2010 − 05/2015 | **Junior Researcher at Institute of High Energy Physics, Vienna**<br>*Member of the CMS experiment at the LHC, CERN*<br>° Research work for Doctoral and Diploma Thesis and a univ. research project<br>° Development and application of complex statistical methods<br>° Project management for physics analyses<br>° Coordination of small working groups, organization and chairing of meetings |

© Springer International Publishing Switzerland 2017

V. Knünz, *Measurement of Quarkonium Polarization to Probe QCD at the LHC*, Springer Theses, DOI 10.1007/978-3-319-49935-2

○ Strategic planning and coordination as 'BPH trigger convener'
○ Co-supervision of bachelor, master and Ph.D. students
○ Preparation of documentation and publications in peer-reviewed journals
○ Presentation of results in several international conferences and seminars

07/2009 – 08/2009 **Laboratory assistant at Atominstitut, Vienna**
*Research project 'Development of method to characterize a PV module'*
○ Study of method to measure the current–voltage curve of a solar module
○ Simulation, measurement and evaluation of uncertainties

06/2008 – 08/2008 **CERN summer-internship, Accelerator-Beams Dep., Geneva**
*Research project 'Simulation of a proton injection scheme into PS2'*
○ Simulation of synchrotron motion, evaluation of feasible injection-schemes

# Education

10/2011 – 03/2015 **Doctoral program in Natural Sciences Technical Physics**
○ Vienna University of Technology (Dr.rer.nat./Ph.D., summa cum laude)
○ *Doctoral Thesis* 'Measurement of Quarkonium Polarization to Probe QCD at the LHC', supervisors: Claudia Wulz, Jožko Strauss

10/2004 – 10/2011 **Diploma program Technical Physics**
○ Vienna University of Technology (Dipl.-Ing./MSc, summa cum laude)
○ *Diploma Thesis* 'Measurement of J/$\psi$ polarization at the CMS experiment', supervisor: Chris Fabjan

09/1995 – 06/2003 **Bundesrealgymnasium Dornbirn Schoren** (Matura, with distinction)

# Internships and Temporary Employment

08/2007 – 09/2007 **Internship at *2XM Millner & Millner* (consulting engineer), Dornbirn**
○ Structural analysis and planning of a climbing garden

07/2006 – 08/2006 **Internship at *KTM Sportmotorcycle AG*, R&D division, Mattighofen**
○ Motorcycle combustion engine studies, thermodynamic simulations

10/2003 – 10/2004 **Zivildienst at child and adolescent psychiatry 'Carina', Feldkirch**
○ Pedagogical care for children with behavioral disorders and disabilities
○ Maintenance and gardening work, transport service

2001 – 2010 **Temporary employment as ski instructor in Lech am Arlberg**
○ Responsibility for groups of up to 15 children
○ Beginners and advanced skiers of all ages

## Awards and Fellowships

○ *CMS Thesis Award 2015*, awarded by the CMS Collaboration Board
○ *Victor-Hess-Award 2015*, awarded by the Particle Physics Division of the Austrian Physical Society
○ *Special Prize for Science 2015*, attributed by the Austrian state Vorarlberg
○ *CERN research fellowship*

## Computer Skills

○ *Programming languages:* C++ ●●●●○, Python ●●●○○, Fortran ●●○○○
○ *Operating systems:* Mac OS ●●●●●, Windows ●●●●○, UNIX ●●●●○
○ *Object-oriented frameworks:* Root, RooFit, CMSSW
○ *Tools:* MS Office, LaTex, Adobe Illustrator, SVN, GitHub, LHC GRID, Eclipse, Matlab, ...

## Language Skills

○ *German:* Mother tongue ●●●●●
○ *English:* Excellent level ●●●●○
○ *French:* Basic level ●○○○○

## Scientific Outreach Activities

○ Exhibit "Confinement in QCD", proposal (FWF), planning, implementation, 2014
○ Article "Head of the month", published by the Austrian HEP community, 10/2012

## Certificates

○ Driver's licences A, B
○ Sailing certificate
○ Snow sport instructor certificate (ski, snowboard)

## Selected Scientific Publications

- CMS Collaboration, '$\Upsilon(nS)$ *polarizations versus particle multiplicity in pp colli-sions at $\sqrt{s} = 7\ TeV$*', PLB 761 (2016) 31
- V. Knünz, '*Measurement of Quarkonium Polarization to Probe QCD at the LHC*', Doctoral Thesis, CERN-THESIS-2015-024, Vienna University of Technology, 2015
- CMS Collaboration, '*Measurement of prompt $J/\psi$ and $\psi(2S)$ double-differential cross sections in pp collisions at $\sqrt{s} = 7\ TeV$*', PRL 114 (2015) 191802
- CMS Collaboration, '*Measurement of the production cross section ratio $\sigma(\chi_{b2}(1P))/\sigma(\chi_{b1}(1P))$ in pp collisions at $\sqrt{s} = 8\ TeV$*', PLB 743 (2015) 383
- V. Knünz for the CMS Collaboration, '*Quarkonium production and polarization in pp collisions with CMS*', PoS (Beauty2014) 012
- C. Lourenço et al., '*A change of perspective in quarkonium production: All data are equal, but some are more equal than others*', Proc. HP2013, NPA 932 (2014) 466
- P. Faccioli, V. Knünz, C. Lourenço, J. Seixas, H. Wöhri, '*Quarkonium production in the LHC era: a polarized perspective*', PLB 736 (2014) 98
- CMS Collaboration, '*Measurement of the prompt $J/\psi$ and $\psi(2S)$ polarizations in pp collisions at $\sqrt{s} = 7\ TeV$*', PLB 727 (2013) 381
- CMS Collaboration, '*Measurement of the $\Upsilon(1S)$, $\Upsilon(2S)$ and $\Upsilon(3S)$ polarizations in pp collisions at $\sqrt{s} = 7\ TeV$*', PRL 110 (2013) 081802
- V. Knünz for the CMS Collaboration, '*Measurement of $\Upsilon(1S)$, $\Upsilon(2S)$ and $\Upsilon(3S)$ polarizations in pp collisions at $\sqrt{s} = 7\ TeV$ with the CMS experiment*', PoS (ICHEP2012) 309
- CMS Collaboration, '*Observation of a new $\Xi_b$ baryon*', PRL 108 (2012) 252002
- V. Knünz, '*Measurement of $J/\psi$ Polarization with the CMS Experiment in Proton–Proton Collisions at $\sqrt{s} = 7\ TeV$*', Diploma Thesis, HEPHY, Vienna University of Technology, 2011
- M. Benedikt, C. Carli, S. Hancock, V. Knünz and I. Vonderhaid, '*Longitudinal painting schemes for H$^-$ charge exchange injection into the PS2*', Proc. of PAC09

On author list of **373** further peer-reviewed publications, as member of the CMS Collaboration.

# Selected Contributions to Conferences and Seminars

- *'A data-driven interpretation of heavy quarkonium measurements at the LHC'*, P. Faccioli et al., QCD@LHC2016, Zurich, Switzerland, Aug. 2016
- *'Measurement of Quarkonium Polarization to Probe QCD at the LHC'*, V. Geisler-Knünz, CMS 2015 Ph.D. thesis award presentation, CMS week plenary, CERN, June 2016
- *'Measurement of quarkonium polarization to probe QCD at the LHC: From puzzles to understanding'*, V. Knünz, Victor Hess award presentation, Vienna, Austria, Sept. 2015
- *'Production of P-wave quarkonia in pp collisions at 7 or 8 TeV at CMS'*, V. Knünz for the CMS Collaboration, QWG2014, CERN, Geneva, Switzerland, Nov. 2014
- *'Quarkonium production in the LHC era: From puzzles to understanding'*, V. Knünz, Experimental Particle Physics Seminars, University of Edinburgh, Scotland, Oct. 2014
- *'Quarkonium production and polarization in pp collisions with the CMS detector'*, V. Knünz for the CMS Collaboration, BEAUTY2014, Edinburgh, Scotland, July 2014
- *'Puzzles, achievements and perspectives in quarkonium production studies'*, P. Faccioli et al., SGW2013, Nantes, France, Dec. 2013
- *'Towards the solution of the "quarkonium polarization puzzle"'*, C. Lourenço et al., HP2013, Stellenbosch, South Africa, Nov. 2013
- *'Quarkonium production and polarization in pp collisions with CMS'*, V. Knünz for the CMS Collaboration, HP2013, Stellenbosch, South Africa, Nov. 2013
- *'Measurement of quarkonium polarization at CMS'*, V. Knünz for the CMS Collaboration, Joint Annual Meeting of ÖPG and SPG, Linz, Austria, Sept. 2013
- *'Higgs searches at the LHC − observation of a new boson with a mass of 125 GeV'*, I. Krätschmer and V. Knünz, Theory Lunch Sem., Univ. of Vienna, Austria, Jan. 2013
- *'Measurement of the $\Upsilon(nS)$ polarizations in pp collisions with the CMS experiment'*, V. Knünz for the CMS Collaboration, ICHEP2012, Melbourne, Australia, July 2012
- *'Prospects for quarkonium polarization measurements at the LHC'*, V. Knünz et al., $8^{th}$ Vienna Central European Seminar on Particle Physics and QFT, Vienna, Austria, Nov. 2011
- *'Measurement of J/$\psi$ production at CMS'*, V. Knünz for the CMS Collaboration, PLHC2011, Perugia, Italy, June 2011
- *'J/$\psi$ polarization scenarios from the Tevatron to the LHC'*, V. Knünz et al., Quarkonium Production Workshop, Vienna, Austria, April 2011
- *'Measurement of quarkonium production at CMS'*, V. Knünz, Annual Meeting of the Austrian Physical Society, Salzburg, Austria, Sept. 2010

# Main or Co-author of the Following CMS Analysis Notes

- *'Prompt $\chi_{c1}$ and $\chi_{c2}$ polarizations in pp collisions at $\sqrt{s} = 8\,TeV$'*, AN-2014-133, in preparation
- *'Low $p_T$ muon and dimuon efficiencies in the 2012 data'*, AN-2014-132, in preparation
- *'$\Upsilon(nS)$ polarizations in pp collisions vs. charged particle multiplicity'*, AN-2015-024, June 2015
- *'Measurement of the $\chi_{cJ}$ prompt yields'*, AN-2014-131, March 2015
- *'Prompt $J/\psi$ and $\psi(2S)$ differential cross sections at $\sqrt{s} = 7\,TeV$'*, AN-2014-003, Feb. 2014
- *'Measurement of the relative production rate of $\chi_{b2}$ and $\chi_{b1}$'*, AN-2013-194, Oct. 2013
- *'$J/\psi$ and $\psi(2S)$ polarizations in pp collisions at $\sqrt{s} = 7\,TeV$'*, AN-2013-016, June 2013
- *'Measurement of the $\Upsilon(nS)$ polarizations at $\sqrt{s} = 7\,TeV$'*, AN-2012-140, June 2012
- *'A new procedure for the determination of angular distribution parameters in dilepton vector meson decays'*, AN-2011-535, June 2012
- *'Low $p_T$ muon and dimuon efficiencies'*, AN-2012-088, May 2012
- *'Measurement of the $\chi_b(3P)$ mass in pp collisions at $\sqrt{s} = 7\,TeV$'*, AN-2012-069, April 2012
- *'Search for a new $\Xi_b$ baryon'*, AN-2012-042, April 2012
- *'Spin alignment of $J/\psi$ mesons in pp collisions at $\sqrt{s} = 7\,TeV$'*, AN-2011-091, March 2011
- *'Studies of the $J/\psi$ polarization fit with a toy Monte Carlo program'*, AN-2011-087, March 2011

Printed in the United States
By Bookmasters